Lecture Notes in Mathematics 1598

Editors:
A. Dold, Heidelberg
F. Takens, Groningen

Jürgen Berndt Franco Tricerri
Lieven Vanhecke

Generalized Heisenberg Groups and Damek-Ricci Harmonic Spaces

Springer

Authors

Jürgen Berndt
Mathematisches Institut
Universität zu Köln
Weyertal 86-90
D-50931 Köln, Germany
E-mail: berndt@mi.uni-koeln.de

Franco Tricerri †
formerly:
Dipartimento di Matematica "U. Dini"
Università di Firenze

Lieven Vanhecke
Department of Mathematics
Katholieke Universiteit Leuven
Celestijnenlaan 200 B
B-3001 Leuven, Belgium
E-mail: fgaga03@cc1.kuleuven.ac.be

Mathematics Subject Classification (1991): 53C20, 53C25, 53C30, 53C40, 22E25

ISBN 3-540-59001-3 Springer-Verlag Berlin Heidelberg New York

CIP-Data applied for

Printed in Germany

Typesetting: Camera-ready TEX output provided by the authors
SPIN: 10130255 46/3142-543210 - Printed on acid-free paper

Preface

The fundamental conjecture about harmonic manifolds has been a source of intensive research during the past decades. Curvature theory plays a fundamental role in this field and is intimately related to the study of the Jacobi operator and its role in the geometry of geodesic symmetries and reflections on a Riemannian manifold.

Our research about harmonic manifolds led in a natural way to the study of spaces with volume-preserving geodesic symmetries and several related classes of manifolds, in particular commutative spaces and Riemannian manifolds all of whose geodesics are orbits of one-parameter groups of isometries. It was also a part of our motivation for developing the theory of homogeneous structures. In this work, the classical and the generalized Heisenberg groups provided a rich collection of examples and counterexamples. It is also well-known that the latter ones take a nice and important place in the florishing research about nilpotent Lie groups and nilmanifolds.

Recently the picture has changed drastically on the one hand by the positive results of Z.I. Szabó and on the other hand by the discovery of the Damek-Ricci harmonic spaces which are the first counterexamples to the fundamental conjecture. These manifolds are Lie groups whose Lie algebras are solvable extensions of generalized Heisenberg algebras. The discovery of these spaces led to a renewed interest in the field, in particular because, just as in the case of the generalized Heisenberg groups, they were found during the work in harmonic analysis and not much attention was given to the detailed study of their geometry and the properties of their curvature as reflected in those of the Jacobi operator.

These notes present a more detailed treatment of this aspect for both classes of manifolds. We do this by relating our study to the several classes of Riemannian manifolds which we have introduced or studied recently in the field of the geometry of the Jacobi operator. It will be shown that they have a rich geometry and provide again answers, examples and counterexamples for several other conjectures and open problems. It is our hope that these notes will stimulate further fruitful research in this area.

During our work in this field, many friends, collaborators and colleagues have contributed by means of their lectures, discussions, joint work, encouragement and interest. They all made this result possible. We are very grateful for their help and for sharing with us their interest and love for mathematics and in particular for geometry. In particular, we thank O. Kowalski, F. Prüfer and F. Ricci.

We also take the opportunity to thank our respective universities, the Consiglio Nazionale delle Richerche (Italy) and the National Fund for Scientific Research (Belgium) for their continued financial support.

Finally we express our deep gratitude to our families for giving us the time needed to do what we enjoy so much.

Köln, Firenze, Leuven
May 1994

Jürgen Berndt, Franco Tricerri, Lieven Vanhecke

To our deep sorrow Franco Tricerri, his wife and his two children died in an airplane crash two weeks after completion of this manuscript. Our loss is immeasurable.

Jürgen Berndt and Lieven Vanhecke

Contents

Chapter 1

Introduction

Nilpotent and solvable Lie groups with left-invariant Riemannian metrics play a remarkable role in Riemannian geometry. At many occasions they arise quite naturally. For example, they appear in the Iwasawa decomposition of the isometry group of a non-compact Riemannian symmetric space. Also, every connected homogeneous Riemannian manifold of non-positive sectional curvature can be represented as a connected solvable Lie group with a left-invariant metric.

Among the nilpotent Lie groups the two-step ones are of particular significance. Some aspects of the geometry of the latter ones, equipped with a left-invariant Riemannian metric, have been treated recently in [Ebe2] and [Ebe3]. The generalized Heisenberg groups form a subclass of the simply connected two-step nilpotent Lie groups with a left-invariant Riemannian metric and, as mentioned by P. Eberlein [Ebe2], should be regarded as the model spaces among these two-step nilpotent Lie groups in a similar way as the Riemannian symmetric spaces among all Riemannian manifolds.

Generalized Heisenberg groups were introduced by A. Kaplan [Kap1] around 1980 in the framework of his research about hypoelliptic partial differential equations. Starting from a composition of quadratic forms, Kaplan defined a class of simply connected two-step nilpotent Lie groups with a left-invariant Riemannian metric which includes the classical Heisenberg groups. The composition of quadratic forms is intimately related to the theory of Clifford modules. In fact, to each representation of the Clifford algebra of $\mathbb{R}^m$ with respect to a negative definite quadratic form one can associate a generalized Heisenberg group with m-dimensional center. This assignment is one-to-one if $m \not\equiv 3 (\operatorname{mod} 4)$. When $m \equiv 3 (\operatorname{mod} 4)$, non-equivalent Clifford modules may yield isometric generalized Heisenberg groups. Nevertheless, using representation theory of Clifford algebras, a complete classification of generalized Heisenberg groups can be achieved.

Both the harmonic analysis and the geometry of the three-dimensional Heisenberg group and its immediate higher-dimensional generalizations have been fruitful fields for research in the past. Since their introduction by A. Kaplan many mathematicians were attracted by the generalized Heisenberg groups in relation with these two fields. See the reference list in [DaRi2]. One of the first remarkable results concerning the geometry of generalized Heisenberg groups was that they are D'Atri spaces, that is, have volume-preserving geodesic symmetries (up to sign). On the other hand, these groups are not naturally reductive as a Riemannian homogeneous space unless the dimension m of the center is one or three. This answered negatively the question whether a D'Atri space is always locally isometric to a naturally reductive Riemannian homogeneous space or not. Moreover, nilmanifolds arising as compact quotients from generalized Heisenberg groups have attracted considerable attention in spectral geometry. For more details see [Gor2], where the author provides, by using suitable compact quotients of generalized Heisenberg groups, the first known examples of closed isospectral Riemannian manifolds which are not locally isometric to each other.

As will be shown in Chapter 3 of these notes, generalized Heisenberg groups also provide examples and counterexamples for other questions and conjectures. But up to now, a systematic study of the geometry of these groups, in particular the aspects relating to the Jacobi operator, was not available. One of the purposes of these notes is to provide a thorough treatment of these aspects based on the explicit research about the spectral theory of this operator and the explicit computation of the Jacobi vector fields vanishing at a point. This method of attack does not only give new geometrical properties but also yields new and more geometrical proofs of known results. Moreover, by doing this, we will relate our research to the different classes of Riemannian manifolds which have been introduced recently in the framework of the study of the geometry of the Jacobi operator. Chapter 2 contains a short survey about these classes including their definitions, known classifications, various characterizations and relations between them.

Our interest in the treatment of D'Atri spaces, as introduced in [Dat], [DaNi1], and [DaNi2], came from the research about harmonic spaces. The fundamental conjecture about harmonic spaces (also referred to as the conjecture of Lichnerowicz) stated that every Riemannian harmonic manifold is locally isometric to a two-point homogeneous space. It was shown that the condition of harmonicity is equivalent to two infinite series of conditions on the curvature tensor and its covariant derivatives, known as the even and odd Ledger conditions. The D'Atri property is equivalent to the set of odd Ledger conditions. Only during the past five years there was a breakthrough in this field on the one hand by the positive results by Z.I. Szabó (see 2.6) and on the other hand by the negative one by E. Damek and F. Ricci. More precisely, the last two authors showed in [DaRi1] that this conjecture is false by proving that there exist suitable extensions of arbitrary generalized Heisenberg groups which are harmonic. Any such extension is a simply connected solvable Lie group with a left-invariant Riemannian metric. Among these Lie groups are the complex hyperbolic spaces, the quaternionic hyperbolic spaces and the Cayley hyperbolic plane. Their horospheres provide realizations of the Heisenberg groups in the complex case and of suitable generalized Heisenberg groups with three- and seven-dimensional center, respectively, in the two other cases. In each of these particular cases the corresponding classical or generalized Heisenberg group is precisely the nilpotent part in the Iwasawa decomposition of the isometry group of the hyperbolic space. The above mentioned extension is then the solvable part in the Iwasawa decomposition and, as a group, is a semi-direct product of the nilpotent group and the real numbers. By imitating this construction of the hyperbolic spaces as solvable Lie groups one obtains from each generalized Heisenberg group a solvable Lie group with a left-invariant Riemannian metric. These particular extensions have been called Damek-Ricci spaces. Any of these spaces is a Hadamard manifold with the corresponding generalized Heisenberg group embedded as a horosphere, and is either one of the above hyperbolic spaces or is non-symmetric. In the latter case each one provides a counterexample to the fundamental conjecture about harmonic spaces. Moreover, as is mentioned in [Gor2], the study of the closed geodesic balls in Damek-Ricci spaces by Z.I. Szabó yielded the first examples of closed isospectral Riemannian manifolds with boundary which are not locally isometric to each other. As concerns the harmonic analysis on the Damek-Ricci spaces we again refer to [DaRi2].

But also here, a detailed study of the geometry of the Damek-Ricci spaces is appealing. Some aspects of it have already been considered by several authors (for more details see Chapter 4). Using again the Jacobi operator, in Chapter 4 we will consider those aspects which are related to some of the classes of manifolds considered in Chapter 2. This leads to several new geometrical characterizations of the symmetric Damek-Ricci spaces. It will also be proved that the Damek-Ricci spaces provide, as in the case for generalized Heisenberg groups, examples and counterexamples to open questions and conjectures. All this gives support for the belief that a further study of their geometry will lead to the discovery of other nice geometrical properties.

A more detailed description of the contents of each chapter will be given at the beginning of each of the respective chapters.

Chapter 2

Symmetric-like Riemannian manifolds

In this chapter we provide some basic material about various classes of Riemannian manifolds which may be regarded as generalizations of Riemannian (locally) symmetric spaces. Our list of such generalizations is not exhaustive. For example, we do not talk about the class of k-symmetric spaces [Kow1] which are natural generalizations of symmetric spaces too. Our selection contains only those spaces which are related to our research on generalized Heisenberg groups and their Damek-Ricci harmonic extensions. Concerning the material about the spaces presented here we have tried to be rather complete as regards known classifications and characterizations. The basic references given here will guide the reader to further results and details on these spaces. See also [Van2] for a selection.

All manifolds are supposed to be connected and of class C^∞. Our sign convention for the Riemannian curvature tensor R is given by $R(X,Y) = [\nabla_X, \nabla_Y] - \nabla_{[X,Y]}$ for all tangent vector fields X, Y, where ∇ denotes the Levi Civita connection.

2.1 Naturally reductive Riemannian homogeneous spaces

Let $M = G/H$ be a Riemannian homogeneous space endowed with a G-invariant Riemannian metric g. The Lie group G is supposed to be connected and to act effectively on M. A decomposition of the Lie algebra $\mathfrak{g}$ of G into $\mathfrak{g} = \mathfrak{h} \oplus \mathfrak{m}$, where $\mathfrak{h}$ is the Lie algebra of H, is said to be *reductive* if $Ad(H)\mathfrak{m} \subset \mathfrak{m}$. If H is connected, a decomposition $\mathfrak{g} = \mathfrak{h} \oplus \mathfrak{m}$ is reductive if and only if $[\mathfrak{h},\mathfrak{m}] \subset \mathfrak{m}$. Note that in the present situation there always exists a reductive decomposition. For $X, Y \in \mathfrak{m}$ we denote by $[X,Y]_\mathfrak{m}$ the projection of $[X,Y]$ onto $\mathfrak{m}$. Each $X \in \mathfrak{g}$ generates a one-parameter subgroup of the group $I(M)$ of isometries of M via $p \mapsto (\exp tX) \cdot p$ and hence induces a Killing vector field X^* on M. If $\mathfrak{g} = \mathfrak{h} \oplus \mathfrak{m}$ is a reductive decomposition of $\mathfrak{g}$, the *natural torsion-free connection* $\bar{\nabla}$ with respect to this decomposition is defined by

$$(\tilde{\nabla}_{X^*}Y^*)_o = \frac{1}{2}[X^*, Y^*]_o = -\frac{1}{2}[X, Y]_{\mathfrak{m}}$$

for all $X, Y \in \mathfrak{m}$, where $\pi(H) = o$ for $\pi : G \to G/H$. Finally, a *homogeneous structure* on M is a tensor field T of type (1,2) such that

$$\tilde{\nabla}g = \tilde{\nabla}R = \tilde{\nabla}T = 0$$

for $\tilde{\nabla} := \nabla - T$, where ∇ is the Levi Civita connection of (M, g) and R the corresponding Riemannian curvature tensor. Then we have the following characterizations (or definitions) of naturally reductive Riemannian homogeneous spaces (for (i) and (ii) see [KoNo,Chapter X,3]; for (iii) see [AmSi, Theorem 5.4] and [TrVa1, Theorem 6.2 and the subsequent remark]).

Proposition 1 [KoNo], [AmSi], [TrVa1] *Let (M, g) be a homogeneous Riemannian manifold. Then (M, g) is a naturally reductive Riemannian homogeneous space if and only if there exist a connected Lie subgroup G of $I(M)$ acting transitively and effectively on M and a reductive decomposition $\mathfrak{g} = \mathfrak{h} \oplus \mathfrak{m}$ of $\mathfrak{g}$, where $\mathfrak{h}$ is the Lie algebra of the isotropy group H at some point in M, such that one of the following equivalent statements holds:*

(i) $g([X, Z]_{\mathfrak{m}}, Y) + g(X, [Z, Y]_{\mathfrak{m}}) = 0$ *for all* $X, Y, Z \in \mathfrak{m}$;

(ii) the Levi Civita connection of (M, g) and the natural torsion-free connection with respect to the decomposition are the same;

(iii) every geodesic in M is the orbit of a one-parameter subgroup of $I(M)$ generated by some $X \in \mathfrak{m}$.

An important observation is that a Riemannian homogeneous space $M = G/H$ might be naturally reductive although for any reductive decomposition $\mathfrak{g} = \mathfrak{h} \oplus \mathfrak{m}$ of $\mathfrak{g}$ none of the statements in the proposition holds. The point is that there might exist another appropriate subgroup $\tilde{G}$ of $I(M)$ such that $M = \tilde{G}/\tilde{H}$ and with respect to which a reductive decomposition satisfies the required conditions. Because of this ambiguity the following result has been proved worthwhile for verifying that certain Riemannian homogeneous spaces are naturally reductive without knowing their isometry group and its transitive subgroups explicitly (see [BeVa4], [BlVa], [GoGoVa1], [GoGoVa2], [Nag], [ToVa], [TrVa1], [TrVa2] for applications).

Proposition 2 [TrVa1] *Let (M, g) be a complete and simply connected Riemannian manifold. Then (M, g) is a naturally reductive Riemannian homogeneous space if and only if there exists a homogeneous structure T on M with $T_v v = 0$ for all tangent vectors v of M.*

Every Riemannian symmetric space is naturally reductive. As the classification of Riemannian symmetric spaces is known since the work of E. Cartan, we concentrate now on non-symmetric naturally reductive spaces. For dimension two the situation is clear since any Riemannian homogeneous space obviously has constant curvature and hence is a locally symmetric space. For non-symmetric naturally reductive Riemannian homogeneous spaces in dimensions three, four and five there are the following results (for dimension three see [TrVa1, Theorem 6.5] and in a more explicit way [Kow3]; for the geometric realizations see [BeVa4]).

Theorem 1 [TrVa1], [Kow3], [BeVa4] *Let (M, g) be a three-dimensional simply connected Riemannian manifold. Then (M, g) is a non-symmetric naturally reductive Riemannian homogeneous space if and only if it is one of the following spaces:*

(i) the Lie group $SU(2)$ with some special left-invariant Riemannian metric g. There is a two-parameter family of left-invariant Riemannian metrics on $SU(2)$ making it into a naturally reductive Riemannian homogeneous space. These metrics are precisely those obtained by considering $SU(2) \approx S^3$ as a geodesic sphere in some two-dimensional complex projective or hyperbolic space equipped with some Fubini-Study metric of constant holomorphic sectional curvature;

(ii) the Lie group $\widetilde{SL(2, \mathbb{R})}$ with some special left-invariant Riemannian metric g. There is a two-parameter family of left-invariant Riemannian metrics on $\widetilde{SL(2, \mathbb{R})}$ making it into a naturally reductive Riemannian homogeneous space. These metrics are precisely those obtained by taking the universal covering of any tube around a one-dimensional complex hyperbolic space embedded totally geodesically in a two-dimensional complex hyperbolic space equipped with some Fubini-Study metric of constant holomorphic sectional curvature. In explicit form, these spaces are given by $M = \mathbb{R}^3[t, x, y]$ with

$$ds^2 = \frac{1}{|a+b|}dt^2 + |a+b|e^{-2t}dx^2 + (dy + \sqrt{2b}e^{-t}dx)^2,$$

where $a, b \in \mathbb{R}$ with $b > 0$ and $a + b < 0$. Geometrically, a and b are the eigenvalues of the Ricci tensor of M, the first one with multiplicity two;

(iii) the three-dimensional Heisenberg group H_3 with any left-invariant Riemannian metric g. There is a one-parameter family of such metrics on H_3 and they are obtained by realizing H_3 as a horosphere in some two-dimensional complex hyperbolic space equipped with some Fubini-Study metric of constant holomorphic sectional curvature. Explicitly, $M = \mathbb{R}^3[x, y, z]$ with

$$ds^2 = \frac{1}{2b}(dx^2 + dz^2 + (dy - x dz)^2),$$

where $b \in \mathbb{R}_+$. Here, $-b$ and b are the eigenvalues of the Ricci tensor of M, the first one with multiplicity two.

Theorem 2 [KoVa1] *Let (M, g) be a four-dimensional simply connected Riemannian manifold. Then (M, g) is a non-symmetric naturally reductive Riemannian homogeneous space if and only if it is isometric to some Riemannian product*

$$SU(2) \times \mathbb{R}\ ,\ \ \widetilde{SL(2, \mathbb{R})} \times \mathbb{R}\ ,\ \ H_3 \times \mathbb{R}\ ,$$

where the first factor is equipped with a naturally reductive Riemannian metric according to the classification in dimension three.

Theorem 3 [KoVa5] *Every five-dimensional simply connected non-symmetric naturally reductive Riemannian homogeneous space is either a Riemannian product $M_1 \times M_2$, where M_1 is $SU(2)$, $\widetilde{SL(2,\mathbb{R})}$ or H_3 with some naturally reductive metric and M_2 is some standard space of constant curvature, or locally isometric to one of the following spaces:*

(i) $(SO(3) \times SO(3))/SO(2)_r$ *or* $(SO(3) \times SL(2,\mathbb{R}))/SO(2)_r$ *or* $(SL(2,\mathbb{R}) \times SL(2,\mathbb{R}))/SO(2)_r$, *where* $SO(2)_r$ *denotes the subgroup consisting of pairs of matrices of the form*

$$\begin{pmatrix} \cos t & -\sin t & 0 \\ \sin t & \cos t & 0 \\ 0 & 0 & 1 \end{pmatrix} \times \begin{pmatrix} \cos rt & -\sin rt & 0 \\ \sin rt & \cos rt & 0 \\ 0 & 0 & 1 \end{pmatrix} \quad (t \in \mathbb{R})$$

and r is a rational number. On each of these spaces there is a family of naturally reductive invariant Riemannian metrics depending on two real parameters. For each of the three types the whole family of locally non-isometric spaces depends on two real parameters and one rational parameter;

(ii) $(H_3 \times SO(3))/SO(2)^{(r)}$ *or* $(H_3 \times SL(2,\mathbb{R}))/SO(2)^{(r)}$, *where* $SO(2)^{(r)}$ *denotes the subgroup consisting of all pairs of matrices of the form*

$$\begin{pmatrix} 1 & 0 & t \\ 0 & 1 & 0 \\ 0 & 0 & 1 \end{pmatrix} \times \begin{pmatrix} \cos rt & -\sin rt & 0 \\ \sin rt & \cos rt & 0 \\ 0 & 0 & 1 \end{pmatrix} \quad (t \in \mathbb{R})$$

and r is a rational number. On each of these spaces there is a family of naturally reductive invariant Riemannian metrics depending on two real parameters. For each of the two types the whole family of locally non-isometric spaces depends on two real parameters and one rational parameter;

(iii) the five-dimensional Heisenberg group H_5. The naturally reductive left-invariant Riemannian metrics on H_5 form a two-parameter family. Explicitly, these spaces are $M = \mathbb{R}^5[x,y,z,u,v]$ with

$$ds^2 = \frac{1}{\rho}(du^2 + dx^2) + \frac{1}{\lambda}(dv^2 + dy^2) + (udx + vdy - dz)^2$$

and $\lambda, \rho \in \mathbb{R}_+$;

(iv) $SU(3)/SU(2)$ *or* $SU(1,2)/SU(2)$, *and on each space there is a family of naturally reductive invariant Riemannian metrics depending on two real parameters.*

Geodesic spheres in two-point homogeneous spaces except $\mathrm{Cay}P^2$ and $\mathrm{Cay}H^2$ are naturally reductive Riemannian homogeneous spaces (see [Zil2] and [TrVa2]). Every simply connected η-umbilical hypersurface of a complex space form is naturally reductive [BeVa4]. This has been extended by Nagai [Nag] to the so-called hypersurfaces of type (A) in complex projective spaces and their corresponding ones in complex hyperbolic spaces. Every simply connected φ-symmetric space (that is, a Sasakian manifold with complete characteristic field such that the reflections with respect to the integral curves of that field are global isometries) is naturally reductive [BlVa]. Every simply connected Killing-transversally symmetric space (that is, a space equipped with a complete unit Killing vector field such that the reflections with respect to the flow lines of that field are global isometries) is naturally reductive (see [GoGoVa1] and [GoGoVa2]). Note that each φ-symmetric space is a Killing-transversally symmetric space.

For further results and references on naturally reductive Riemannian homogeneous spaces we refer to J.E. D'Atri and W. Ziller [DaZi], who also classified all naturally reductive compact simple Lie groups. For a treatment of the non-compact semisimple case, see C. Gordon [Gor1].

2.2 Riemannian g.o. spaces

A Riemannian manifold (M, g) is said to be a *Riemannian g.o. space* [KoVa7] if every geodesic in M is the orbit of a one-parameter subgroup of the group of isometries of M. Clearly, any such space is homogeneous. From Proposition 1(iii) in 2.1 we derive immediately

Proposition *Every naturally reductive Riemannian homogeneous space is a Riemannian g.o. space.*

O. Kowalski and the third author [KoVa7] have proved that the converse holds if the dimension is less than six.

Theorem 1 [KoVa7] *Every simply connected Riemannian g.o. space of dimension* ≤ 5 *is a naturally reductive Riemannian homogeneous space.*

Combining this with Theorems 1, 2 and 3 in 2.1 yields a classification of all simply connected Riemannian g.o. spaces of dimension less than six. For dimension six the converse does not hold. In fact, there is the following result:

Theorem 2 [KoVa7] *The following six-dimensional simply connected Riemannian g.o. spaces (and only those) are never naturally reductive:*

(i) (M, g) *is a two-step nilpotent Lie group with two-dimensional center, provided with a left-invariant Riemannian metric such that the maximal connected isotropy group is isomorphic to* $SU(2)$ *or* $U(2)$*. All these Riemannian g.o. spaces depend on three real parameters;*

(ii) $(\tilde{M}, g)$ *is the universal covering space of a homogeneous Riemannian manifold of the form* $M = SO(5)/U(2)$ *or* $M = SO(4,1)/U(2)$*, where* $SO(5)$ *or* $SO(4,1)$ *is the identity component of the full isometry group, respectively. In each case all admissible Riemannian metrics depend on two real parameters.*

Every geodesic sphere in a two-point homogeneous space except CayP^2 or CayH^2 is a Riemannian g.o. space since it is naturally reductive. For CayP^2 and CayH^2 it is still an open problem whether the geodesic spheres are g.o. spaces or not.

2.3 Weakly symmetric spaces

A Riemannian manifold M is said to be a *weakly symmetric space* [Sel] if there exist a subgroup G of the isometry group $I(M)$ of M acting transitively on M and an isometry f of M with $f^2 \in G$ and $fGf^{-1} = G$ such that for all $p, q \in M$ there exists a $g \in G$ with $g(p) = f(q)$ and $g(q) = f(p)$. It can easily be seen that any Riemannian symmetric space is weakly symmetric. There are the following geometrical characterizations of weakly symmetric spaces:

Proposition [BePrVa1], [BeVa5] *Let (M, g) be a Riemannian manifold. Then the following statements are equivalent:*

(i) M is a weakly symmetric space;

(ii) for any two points $p, q \in M$ there exists an isometry of M mapping p to q and q to p;

(iii) for every maximal geodesic γ in M and any point m of γ there exists an isometry of M which is an involution on γ with m as fixed point.

Note that Riemannian manifolds having property (iii) have been introduced by Szabó [Sza2] as *ray symmetric spaces.*

In dimensions three and four the simply connected weakly symmetric spaces are completely classified.

Theorem 1 [BeVa5] *A three- or four-dimensional simply connected Riemannian manifold is a weakly symmetric space if and only if it is a naturally reductive Riemannian homogeneous space (see Theorems* 1 *and* 2 *in* 2.1*).*

We also have the following further examples of non-symmetric weakly symmetric spaces.

Theorem 2 [BeVa5] *Each of the following hypersurfaces, endowed with the induced Riemannian metric of the ambient space, is a weakly symmetric space for $n \geq 2$:*

ambient space	hypersurface
$\mathbb{C}P^n$	*tube around* $\{p\}, \mathbb{C}P^1, \ldots,$ *or* $\mathbb{C}P^{n-1}$
$\mathbb{H}P^n$	*tube around* $\{p\}, \mathbb{H}P^1, \ldots,$ *or* $\mathbb{H}P^{n-1}$
$\mathrm{Cay}P^2$	*tube around* $\{p\}$ *or* $\mathrm{Cay}P^1$
$\mathbb{C}H^n$	*horosphere; tube around* $\{p\}, \mathbb{C}H^1, \ldots,$ *or* $\mathbb{C}H^{n-1}$
$\mathbb{H}H^n$	*horosphere; tube around* $\{p\}, \mathbb{H}H^1, \ldots,$ *or* $\mathbb{H}H^{n-1}$
$\mathrm{Cay}H^2$	*horosphere; tube around* $\{p\}$ *or* $\mathrm{Cay}H^1$.

Other examples, which were discovered only very recently, will be treated in forthcoming papers.

2.4 Commutative spaces

A *commutative space* is a Riemannian homogeneous space whose algebra of all invariant (with respect to the connected component of the full isometry group) differential operators is commutative. I.M. Gelfand [Gel] has proved that any Riemannian symmetric space is commutative (see also [Hel1, p. 396]). This was generalized by A. Selberg to the class of weakly symmetric spaces.

Proposition [Sel] *Every weakly symmetric space is a commutative space.*

Note that it is unknown whether the converse holds.

For dimension less or equal than five the simply connected commutative spaces are well-known:

Theorem [Kow3], [KoVa3], [Bie] *A simply connected Riemannian manifold of dimension* ≤ 5 *is a commutative space if and only if it is a naturally reductive Riemannian homogeneous space (see the theorems in* 2.1*).*

This result cannot be extended to higher dimensions. For example, the Riemannian homogeneous space $SU(3)/T$, where T is a maximal torus in $SU(3)$, provides an example of a six-dimensional naturally reductive Riemannian homogeneous space which is not commutative [Jim1]. In [Jim2] J.A. Jiménez also provides examples of non-commutative naturally reductive Riemannian homogeneous spaces for any odd dimension greater or equal than five. Other examples of this kind are provided by the Stiefel manifolds of orthonormal two-frames in $\mathbb{R}^n$, $n > 30$ [Jim3]. On the other hand, the six-dimensional generalized Heisenberg group with two-dimensional center is a commutative space which is in no way naturally reductive (see Theorems 1 and 3 in 3.2). We do not know whether there are commutative spaces which are not Riemannian g.o. spaces. See also [Hel2] for further results and references about commutative spaces.

2.5 Probabilistic commutative spaces

A *probabilistic commutative space* is a Riemannian manifold M such that $L_s L_t = L_t L_s$ holds locally for all sufficiently small $s, t \in \mathbb{R}_+$. Here, the operator L is the second mean value operator on M defined by

$$L_t f(p) := \frac{1}{\omega_{n-1}} \int_{S^{n-1}(1)} f(\exp_p(t\xi))d\xi \ ,$$

where $f \in C^0(M)$, $p \in M$, $n = \dim M$, $S^{n-1}(1)$ is the unit sphere in T_pM, ω_{n-1} is the volume of $S^{n-1}(1)$, $\exp_p$ is the exponential map of M at p, and $d\xi$ is the volume element of $S^{n-1}(1)$. This concept originated from a work by P.H. Roberts and H.D. Ursell [RoUr] on compact Riemannian manifolds on which any two random steps commute. The general case has then been treated by O. Kowalski and F. Prüfer [Kow2], [KoPr], [Pru].

Proposition 1 [KoPr] [Kow2] *An analytic Riemannian manifold M is probabilistic commutative if and only if all Euclidean Laplacians of order $2k$ ($k \in \mathbb{N}$) commute.*

The Euclidean Laplacians are defined as follows (see for example [GrWi]). Let p be a point in an analytic Riemannian manifold M and $x_1, \ldots, x_n$ normal coordinates on M centered at p. Define a local differential operator $\bar{\Delta}_p$ by

$$\bar{\Delta}_p f := \sum_{i=1}^{n} \frac{\partial^2 f}{\partial x_i^2}$$

and put

$$\tilde{\Delta}^{(k)} f(p) := (\bar{\Delta}_p)^k f(p) .$$

Then $\tilde{\Delta}^{(k)}$ is a global analytic differential operator on M, called the *Euclidean Laplacian of order $2k$*. Each $\tilde{\Delta}^{(k)}$ is a formally self-adjoint operator on the space of all C^∞-functions on M with compact support.

Proposition 1 immediately implies

Proposition 2 *Every commutative space is also probabilistic commutative.*

The converse is not true, as we shall see in 4.5. Classifications of probabilistic commutative spaces are known only for dimension three.

Theorem [Kow3] *A three-dimensional simply connected complete Riemannian manifold is a probabilistic commutative space if and only if it is a naturally reductive Riemannian homogeneous space (see Theorem 1 in 2.1).*

2.6 Harmonic spaces

For a survey on these spaces containing results up to 1982 see [Van1], where the references for the following equivalent definitions or characterizations of harmonic spaces can be found. In particular, we refer to [Bes1] and [RuWaWi].

Proposition 1 *A Riemannian manifold (M, g) is a harmonic space if and only if one the following equivalent statements holds:*

(i) at each point $m \in M$ there exists a normal neighborhood of M on which the Laplace equation $\Delta u = 0$ admits a real non-constant solution depending only upon the distance r to m and being analytic for $r \neq 0$;

(ii) for each point $m \in M$ the volume density function $\theta = (\det g_{ij})^{1/2}$ (in normal coordinates centered at m) is a radial function;

(iii) *for each* $m \in M$ *the function* $\Delta\Omega_m$ *is a function of* Ω_m, *where* $\Omega_m(p) := r^2(p)/2$ *and* $r(p)$ *is the distance from* m *to* p;

(iv) *every sufficiently small geodesic sphere in* M *has constant mean curvature;*

(v) *every sufficiently small geodesic sphere in* M *(where* $\dim M > 2$*) has constant scalar curvature;*

(vi) $M_r f(m) = f(m)$ *for all* $m \in M$, *all sufficiently small* $r \in \mathbb{R}_+$, *and all harmonic functions* f *defined on a neighborhood containing the geodesic ball* $B_m(r)$;

(vii) $M_r f(m) = L_r f(m)$ *for all* $m \in M$, *all sufficiently small* $r \in \mathbb{R}_+$, *and all harmonic functions* f *defined on a neighborhood containing the geodesic ball* $B_m(r)$.

Here, $M_r f(m)$ denotes the first mean value operator on M defined by

$$M_r f(m) = \frac{1}{\mathrm{vol}\, G_m(r)} \int_{G_m(r)} f do\ ,$$

where $G_m(r) = \exp_m(S^{n-1}(r))$ is the geodesic sphere of radius r and with center m, and do denotes the Riemannian volume element of $G_m(r)$.

Using results in [Kow2], [KoPr] or [Sza2] one can prove that

Proposition 2 *Every harmonic space is a probabilistic commutative space.*

We shall see later that the word *probabilistic* in this statement cannot be dropped. In fact, every non-symmetric Damek-Ricci space is a harmonic space which is not commutative (see Corollary 2 in 4.5).

Using characterization (iv) it is easy to see that each two-point homogeneous space is harmonic. For a long time a conjecture on harmonic spaces stated that every harmonic space is locally isometric to a two-point homogeneous space. This conjecture was known to be true for $\dim M \leq 4$, or if the universal covering space of M is compact, or if M is compact and has non-negative scalar curvature (see [Sza1], [Sza2] for details). Moreover, in [Cao] it is proved that a compact harmonic Kähler manifold of negative sectional curvature is isometric to a compact quotient of complex hyperbolic space, $\mathbb{C}H^n/\Gamma$, up to a constant scalar factor. A recent and more general result states that the Riemannian universal covering space $\tilde{M}$ of a compact Riemannian manifold M, such that $\tilde{M}$ has strictly negative sectional curvature and is *asymptotically harmonic* (that is, all its horospheres have constant mean curvature), is a rank one symmetric space [BeCoGa]. Only recently E. Damek and F. Ricci [DaRi1] found the examples of non-symmetric harmonic spaces which are the topic of Chapter 4.

We also want to point out that the Riemannian product of two probabilistic commutative spaces is still commutative, whereas the analogue for harmonic spaces does not hold. So the class of probabilistic commutative spaces is strictly bigger than that formed by the harmonic spaces. Finally we mention that every harmonic space is an Einstein space.

2.7 D'Atri spaces

A Riemannian manifold is said to be a *D'Atri space* [VaWi2] if its local geodesic symmetries are volume-preserving (up to sign). Such spaces have been studied first by J.E. D'Atri and H.K. Nickerson [DaNi1]. Nice characterizations of such spaces in terms of the mean curvature $h_m(p)$ of geodesic spheres $G_m(r)$ with center m and radius r, $p \in G_m(r)$, are given by

Proposition 1 [Dat], [KoVa6], [VaWi1] *For any Riemannian manifold M the following statements are equivalent:*

(i) M is a D'Atri space;

(ii) $h_m(p) = h_p(m)$ for all $m, p \in M$ sufficiently close;

(iii) $h_m(\exp_m(r\xi)) = h_m(\exp_m(-r\xi))$ for all $m \in M$, all unit vectors $\xi \in T_mM$ and $r \in \mathbb{R}_+$ sufficiently small;

(iv) $h_{\exp_m(r\xi)}(m) = h_{\exp_m(-r\xi)}(m)$ for all $m \in M$, all unit vectors $\xi \in T_mM$ and $r \in \mathbb{R}_+$ sufficiently small.

Classifications have been obtained only for dimensions less or equal than three. For dimension two the classification follows immediately from Proposition 5 below, and for dimension three we have

Theorem [Kow3] *A three-dimensional simply connected complete Riemannian manifold is a D'Atri space if and only if it is a naturally reductive Riemannian homogeneous space (see Theorem* 1 *in* 2.1*).*

An open problem is whether any D'Atri space is locally homogeneous. From Proposition 1 in 2.6 and the above proposition it follows that

Proposition 2 *Every harmonic Riemannian manifold is a D'Atri space.*

Further, we have

Proposition 3 [KoVa4] *Every Riemannian g.o. space, and in particular every naturally reductive Riemannian homogeneous space, is a D'Atri space.*

In the special case of naturally reductive Riemannian homogeneous spaces this result was proved by J.E. D'Atri and H.K. Nickerson in [Dat] and [DaNi2]. Next, we have

Proposition 4 [KoPr] *Every probabilistic commutative space, and in particular every commutative space, is a D'Atri space.*

The special case of commutative spaces has been treated by O. Kowalski and the third author in [KoVa2]. Finally, we mention

Proposition 5 *Every D'Atri space is analytic in normal coordinates and has constant scalar curvature.*

In fact, the statement of Proposition 5 holds for the more general class of Riemannian manifolds whose Ricci tensor is a Killing tensor (that is, the cyclic sum over all entries in the covariant derivate of the Ricci tensor vanishes); see [BeVa1] for more details.

Of particular interest is the situation in dimension four. Partial classifications have been obtained by J.T. Cho, K. Sekigawa and the third author [SeVa1], [SeVa2], [ChSeVa]; but up to now there is no complete classification. See also [Bes2], [Van2], [Wil], and [KoPrVa] for further references.

2.8 $\mathfrak{C}$- and $\mathfrak{P}$-spaces

The concept of $\mathfrak{C}$- and $\mathfrak{P}$-spaces is due to the first and third author. A Riemannian manifold M is called a *$\mathfrak{C}$-space* if for every geodesic γ in M the eigenvalues of the associated Jacobi operator field $R_\gamma := R(., \dot\gamma)\dot\gamma$ are constant; and M is called a *$\mathfrak{P}$-space* if for any geodesic γ in M the associated Jacobi operator field R_γ is diagonalizable by a parallel orthonormal frame field along γ. Both classes generalize Riemannian symmetric spaces, for *a Riemannian manifold is locally symmetric if and only if it is both a $\mathfrak{C}$-space and a $\mathfrak{P}$-space* [BeVa1].

A two-dimensional Riemannian manifold is a $\mathfrak{C}$-space if and only if it is of constant curvature. Every two-dimensional Riemannian manifold is a $\mathfrak{P}$-space. For dimension three we have

Theorem 1 [BeVa1] *A three-dimensional complete and simply connected Riemannian manifold is a $\mathfrak{C}$-space if and only if it is a naturally reductive Riemannian homogeneous space (see Theorem 1 in 2.1). If M is a non-complete or a non-simply connected $\mathfrak{C}$-space, it is locally isometric to a complete and simply connected one.*

Theorem 2 [BeVa1] *Let M be a three-dimensional $\mathfrak{P}$-space of class C^∞. Then M is almost everywhere (namely on the set where the number of distinct eigenvalues of the Ricci tensor of M is locally constant; this is an open and dense subset of M) locally isometric to one of the following spaces:*

(i) a space of constant Riemannian sectional curvature;

(ii) a warped product of the form $M_1 \times_f M_2$, where M_1 is a one-dimensional Riemannian manifold, M_2 is a two-dimensional Riemannian manifold, and f is a positive function on M_1;

(iii) a warped product of the form $M_2 \times_f M_1$, where M_1 is a one-dimensional Riemannian manifold, M_2 is a Liouville surface, and f is given by

$$f^2(x,y) = |\varphi(x) \cdot \psi(y)| \ ,$$

where the functions φ and ψ come from a (local) Liouville form

$$(\varphi(x) + \psi(y)) \cdot (dx^2 + dy^2)$$

of the Riemannian metric of M_2;

(iv) a three-dimensional Riemannian manifold with Riemannian metric of the form

$$ds^2 = \mathfrak{S}_{1,2,3}\, F_1(x_1)|x_1 - x_2||x_1 - x_3| dx_1^2 \,,$$

where $\mathfrak{S}$ denotes the cyclic sum and F_1, F_2, F_3 are positive functions.

Every real analytic Riemannian manifold of one of these types is a $\mathfrak{P}$-space.

For higher dimensions we do not know of any classification. Concerning examples of $\mathfrak{C}$-spaces we have

Proposition 1 [BeVa1] *Every Riemannian g.o. space and every commutative space is a $\mathfrak{C}$-space.*

All our known examples of $\mathfrak{C}$-spaces up to now belong to these two classes. We shall see in 3.4 that the generalized Heisenberg groups provide a third class of $\mathfrak{C}$-spaces. All these examples are homogeneous manifolds and it is still unknown whether any $\mathfrak{C}$-space is in fact a locally homogeneous space - in contrast to the $\mathfrak{P}$-spaces where we have a lot of examples which are not locally homogeneous (see also [Boe1], [Boe2], [Cho]).

There are the following characterizations of $\mathfrak{C}$- and $\mathfrak{P}$-spaces.

Proposition 2 [BeVa1], [BeVa2], [BePrVa1] *Let (M, g) be an n-dimensional Riemannian manifold. Then M is a $\mathfrak{C}$-space if and only if one of the following equivalent statements holds:*

(i) for each $p \in M$ and each $v \in T_pM$ there exists a (skew-symmetric) endomorphism T_v of T_pM so that $R'_v := (\nabla_v R)(.,v)v = [R_v, T_v]$ with $R_v := R(.,v)v$;

(ii) for each geodesic γ in M there exists a skew-symmetric tensor field T_γ along γ such that $R'_\gamma = [R_\gamma, T_\gamma]$;

(iii) for each $v \in TM$ and eigenvalue κ of R_v the eigenspace of R_v with respect to κ is mapped by R'_v into its orthogonal complement;

(iv) for each $v \in TM$ and eigenvalue κ of R_v there exists a corresponding eigenvector $w \neq 0$ such that $g(R'_v w, w) = 0$;

(v) for each $m \in M$ every local geodesic symmetry s_m of M at m preserves the eigenvalues of $R_\gamma(r)$, where γ is an arbitrary geodesic in M with $\gamma(0) = m$ and $r \in \mathbb{R}_+$ is sufficiently small;

(vi) for each $p \in M$ and $v \in T_pM$ there exists a skew-symmetric endomorphism T_v of T_pM such that $T_{-v} = -T_v$ and the eigenvalues of $R'_v - [R_v, T_v]$ are independent of v;

(vii) for every geodesic γ in M defined at 0 and every eigenvalue function c of the associated Ledger tensor C we have $c'''(0) = 0$ (note that always $c(0) = 1$, $c'(0) = 0$ and $c''(0) = -2\kappa/3$ with some suitable eigenvalue κ of $R_\gamma(0)$, and the Ledger tensor C is defined by $C(r) := rA(r)$, where $A(r)$ denotes the shape operator at $\gamma(r)$ of the geodesic sphere centered at $\gamma(0)$);

(viii) for every $p \in M$, $v \in T_pM$, eigenvalue κ of R_v and corresponding eigenvector w we have $g_S(R^{TM}_{S(v)}w^H, w^V) = 0$, where g_S is the Sasaki metric on TM, S the geodesic spray of M and ^{H} and ^{V} denote the horizontal and vertical lift with respect to the canonical bundle map $TM \to M$, respectively;

(ix) for every $k \in \{1, \ldots, n-1\}$ the polynomial $P_k : TM \to \mathbb{R}, v \mapsto \operatorname{trace} R_v^k$ is a first integral of the geodesic flow of M;

(x) $P_1 + \ldots + P_{n-1}$ is a first integral of the geodesic flow of M;

(xi) for every $k \in \{1, \ldots, n-1\}$ the symmetric tensor field S_k on M obtained by polarization of P_k is a Killing tensor.

Proposition 3 [BeVa1] *Let (M, g) be an n-dimensional real analytic Riemannian manifold. Then the following statements are equivalent:*

(i) M is a $\mathfrak{P}$-space;

(ii) $R_v \circ R_v' = R_v' \circ R_v$ for all $v \in TM$;

(iii) the basic Jacobi fields on M are of the form as in locally symmetric spaces, that is, they arise from multiplying appropriate parallel vector fields with particular solutions of scalar Jacobi equations (here, a basic Jacobi field Y along a geodesic is a Jacobi field with $Y(t) = 0$ or $Y'(t) = 0$ at some point);

(iv) the principal curvature spaces of any family of (sufficiently small) geodesic spheres in M are invariant with respect to parallel translation along the radial geodesics emanating from the center of this family;

(v) all (sufficiently small) geodesic spheres in M are curvature-adapted, that is, the shape operator and normal Jacobi operator commute.

Note that the symmetric tensor field S_1 in statement (xi) of Proposition 2 is just the Ricci tensor. So the Ricci tensor on any $\mathfrak{C}$-space is a Killing tensor and hence we have (see also the remark after Proposition 5 in 2.7)

Proposition 4 *Every $\mathfrak{C}$-space is analytic in normal coordinates and has constant scalar curvature.*

2.9 $\mathfrak{C}_0$-spaces

The $\mathfrak{C}_0$-spaces form a subclass of the $\mathfrak{C}$-spaces and were introduced in [BePrVa1]. There are the following defining or characterizing properties of these spaces.

Proposition 1 [BePrVa1] *A Riemannian manifold (M, g) is a $\mathfrak{C}_0$-space if and only if one of the following equivalent statements holds:*

(i) for every geodesic γ in M there exists a parallel skew-symmetric tensor field T_γ along γ such that $R_\gamma' = [R_\gamma, T_\gamma]$;

(ii) for every geodesic γ in M defined at 0 there exists a skew-symmetric endomorphism T of $T_{\gamma(0)}M$ such that

$$R_\gamma(t) = e^{-tT} \circ R_\gamma(0) \circ e^{tT} ,$$

where different tangent spaces along γ are identified via parallel translation;

(iii) for every geodesic γ in M defined at 0 there exists a skew-symmetric endomorphism T of $T_{\gamma(0)}M$ such that all the higher order Jacobi operators $R_\gamma^{(k)}$, $k = 0, 1, 2, \ldots$, satisfy

$$R_\gamma^{(k)}(t) = e^{-tT} \circ R_\gamma^{(k)}(0) \circ e^{tT} ,$$

where different tangent spaces along γ are identified via parallel translation.

Characterization (ii) says that $\mathfrak{C}_0$-spaces are precisely those Riemannian manifolds for which the Jacobi operator R_γ along any geodesic γ is obtained by conjugation of the Jacobi operator $R_\gamma(0)$ at a single point with a suitable one-parameter subgroup of the orthogonal group. The space is locally symmetric if and only if this subgroup can always be chosen as the trivial one.

As concerns examples of $\mathfrak{C}_0$-spaces we have

Proposition 2 [BePrVal] *Every Riemannian g.o. space is a $\mathfrak{C}_0$-space.*

This immediately implies

Theorem *A three-dimensional simply connected complete Riemannian manifold is a $\mathfrak{C}_0$-space if and only if it is a naturally reductive Riemannian homogeneous space (see Theorem 1 in 2.1).*

We know that commutative spaces are $\mathfrak{C}$-spaces, but we do not know whether they are also $\mathfrak{C}_0$-spaces. We also do not have a single example of a $\mathfrak{C}$-space which is not a $\mathfrak{C}_0$-space. Another open problem is whether any $\mathfrak{C}$-space is a D'Atri space. For the subclass of all $\mathfrak{C}_0$-spaces this is known to be true.

Proposition 3 [BePrVal] *Every $\mathfrak{C}_0$-space is a D'Atri space.*

According to J.E. D'Atri and H.K. Nickerson [DaNi2] we say that a Riemannian manifold M has a *special curvature tensor* if it admits a tensor field T of type (1,2) satisfying

$$(\nabla_X R)(Y,X)X = T_X R(Y,X)X - R(T_X Y, X)X \quad \text{and} \quad (\nabla_X T)_X = 0 .$$

It follows from Proposition 1(i) that every space with a special curvature tensor is a $\mathfrak{C}_0$-space.

2.10 $\mathfrak{TC}$-spaces

Consider the following geometric configuration. Let γ be a geodesic in a Riemannian manifold M parametrized by arc length and defined on an open interval containing

0. We put $m := \gamma(0)$ and choose $r \in \mathbb{R}_+$ such that $p := \gamma(r)$ and $q := \gamma(-r)$ are defined. If r is sufficiently small, the geodesic spheres $G_p(r)$ and $G_q(r)$ of radius r around p and q, respectively, are smoothly embedded hypersurfaces in M. Further, m lies on both geodesic spheres and $G_p(r)$ and $G_q(r)$ are tangent to each other at m. M is called a *$\mathfrak{TC}$-space* if for any such configuration the principal curvatures (counted with multiplicities) of $G_p(r)$ and $G_q(r)$ are the same. Classifications of $\mathfrak{TC}$-spaces in low dimensions follow from the following result.

Theorem [BeVa3] *Let M be a simply connected complete Riemannian manifold with dimension ≤ 3. Then M is a $\mathfrak{TC}$-space if and only if it is a naturally reductive Riemannian homogeneous space (see Theorem* 1 *in* 2.1*).*

(Note that the assumption of analyticity in [BeVa3] is unnecessary since every $\mathfrak{TC}$-space is a D'Atri space (see below) and hence real analytic in normal coordinates.) Examples of $\mathfrak{TC}$-spaces arise from the following two results.

Proposition 1 [BePrVa1] *Every weakly symmetric space is a $\mathfrak{TC}$-space.*

Proposition 2 [BePrVa1] *Every Riemannian manifold which is locally homothetic to a Sasakian space form is a $\mathfrak{TC}$-space.*

It is an open problem whether naturally reductive Riemannian homogeneous spaces, Riemannian g.o. spaces, $\mathfrak{C}_0$-spaces, $\mathfrak{C}$-spaces or commutative spaces are $\mathfrak{TC}$-spaces. From the classifications we just have the partial result [BePrVa1] that every simply connected naturally reductive Riemannian homogeneous space, Riemannian g.o. space or commutative space of dimension ≤ 4 is a $\mathfrak{TC}$-space. Consequences of the $\mathfrak{TC}$-property are the $\mathfrak{C}$- and the D'Atri-property.

Proposition 3 [BePrVa1] *Every $\mathfrak{TC}$-space is a $\mathfrak{C}$-space.*

Proposition 4 [BeVa3] *Every $\mathfrak{TC}$-space is a D'Atri space.*

The motivation for studying $\mathfrak{TC}$-spaces emerged from a result by the third author and T.J. Willmore [VaWi2] stating that a Riemannian manifold is locally symmetric if and only if for any configuration of geodesic spheres (as described in the beginning) the shape operators of $G_p(r)$ and $G_q(r)$ coincide at m. The last condition means that the two shape operators have the same eigenvalues and are simultaneously diagonalizable. It is natural to split up here the two conditions on the shape operators, the first one leading to the $\mathfrak{TC}$-spaces and the second one to the so-called *$\mathfrak{TP}$-spaces*. In [BeVa3] it was proved that in fact the classes $\mathfrak{P}$ and $\mathfrak{TP}$ coincide in the real analytic case.

2.11 $\mathfrak{SC}$-spaces

A Riemannian manifold (M, g) is said to be an *$\mathfrak{SC}$-space* if small geodesic spheres in M have the same principal curvatures (counted with multiplicities) at antipodal points. Concerning $\mathfrak{SC}$-spaces we have similar results as for $\mathfrak{TC}$-spaces.

Theorem [BePrVa1] *Let M be a simply connected complete Riemannian manifold of dimension ≤ 3. Then M is an $\mathfrak{S}\mathfrak{C}$-space if and only if it is a naturally reductive Riemannian homogeneous space (see Theorem 1 in 2.1).*

Proposition 1 [BePrVa1] [BeVa5] *Every weakly symmetric space is an $\mathfrak{S}\mathfrak{C}$-space.*

Proposition 2 [BePrVa1] *Every Riemannian manifold which is locally homothetic to a Sasakian space form is an $\mathfrak{S}\mathfrak{C}$-space.*

Proposition 3 [BePrVa1] *Every $\mathfrak{S}\mathfrak{C}$-space is a $\mathfrak{C}$-space.*

Proposition 4 [BePrVa1] *Every $\mathfrak{S}\mathfrak{C}$-space is a D'Atri space.*

It is not yet clear whether the classes of $\mathfrak{T}\mathfrak{C}$- and $\mathfrak{S}\mathfrak{C}$-spaces coincide or not. A partial answer is given by

Proposition 5 [BePrVa1] *Let M be a $\mathfrak{C}_0$-space or a commutative space. Then M is a $\mathfrak{T}\mathfrak{C}$-space if and only if it is an $\mathfrak{S}\mathfrak{C}$-space.*

2.12 Osserman spaces

The notion of an Osserman space is due to the

Osserman conjecture [Oss] *Every Riemannian manifold with globally constant eigenvalues for the Jacobi operators is locally isometric to a two-point homogeneous space.*

Note that in the conjecture only Jacobi operators with respect to *unit* tangent vectors are considered. It is easy to see that each two-point homogeneous space has the property stated in the conjecture. A Riemannian manifold with globally constant eigenvalues for the Jacobi operators is called a *globally Osserman space* [GiSwVa]. Up to now the conjecture is known to be true in the following cases:

Theorem [Chi1], [Chi2], [Chi3] *Let M be an n-dimensional globally Osserman space. Then M is locally isometric to a two-point homogeneous space in the following cases:*

(i) n *is odd;*

(ii) $n \equiv 2 (\mathrm{mod}\ 4)$;

(iii) $n = 4$;

(iv) $n = 4k$, $k > 1$, *and M is a simply connected compact quaternionic Kähler manifold with vanishing second Betti number;*

(v) M is a Kähler manifold of non-negative or non-positive sectional curvature;

(vi) M satisfies the following axioms:

(1) R_v *has precisely two different constant eigenvalues independent of* $v \in SM$ *(the unit sphere bundle in* TM*);*

(2) *let* λ *and* μ *be the two eigenvalues and for* $v \in SM$ *denote by* $E_\mu(v)$ *the span of* v *and the eigenspace of* R_v *with eigenvalue* μ*; then* $E_\mu(w) = E_\mu(v)$ *whenever* $w \in E_\mu(v)$.

One might also consider a local version, namely that for each $p \in M$ the eigenvalues of the Jacobi operator are independent of unit vectors $v \in T_pM$ (but may vary with $p \in M$). Such spaces are called *pointwise Osserman spaces*. Surprisingly there are pointwise Osserman spaces which are not globally Osserman spaces. For example, four-dimensional self-dual Einstein manifolds are always pointwise Osserman spaces, but not locally isometric to a two-point homogeneous space in general. See [GiSwVa] for more details in the local situation.

Chapter 3

Generalized Heisenberg groups

In Section 1 we provide basic material about generalized Heisenberg groups, most of which has been presented previously in [Kap1], [Kap2], [Kap3] and [Rie1]. In Section 2 we summarize the known results, which are due to [Kap3], [Ric] and [Rie2], about the question which of the generalized Heisenberg groups belong to some of the classes of Riemannian manifolds discussed in Chapter 2. We also add a new result concerning weakly symmetric spaces. In Section 3 we determine the spectrum of the Jacobi operator and the corresponding eigenspaces. From this we conclude that none of the generalized Heisenberg groups is a pointwise Osserman space. The explicit knowledge of these data enables us to provide in Section 4 two alternative proofs for the fact that every generalized Heisenberg group is a $\mathfrak{C}$-space. Consequently, none of the generalized Heisenberg groups is a $\mathfrak{P}$-space. In Section 5 we show that every generalized Heisenberg group is even a $\mathfrak{C}_0$-space. The explicit expression of the tensor field T there is perhaps the crucial point in these notes, as without it the results in the subsequent sections would hardly be obtained. In Section 6 we draw some further conclusions from the results which have been obtained so far. In Section 7 we determine all Jacobi fields in generalized Heisenberg groups which vanish at the identity. The method introduced here for doing this generalizes the one used by I. Chavel [Cha] for normal homogeneous Riemannian manifolds and by W. Ziller [Zil1] for naturally reductive Riemannian homogeneous spaces. The subsequent sections then contain applications of the explicit knowledge of these Jacobi fields. In Section 8 we compute some conjugate points in generalized Heisenberg groups. For the Heisenberg groups we obtain a complete classification of the conjugate points. The shape operator of geodesic spheres in generalized Heisenberg groups is the topic of Section 9. From the structure of the Jacobi fields we conclude that the principal curvatures of small geodesic spheres in generalized Heisenberg groups are the same at antipodal points. This means that every generalized Heisenberg group is an $\mathfrak{SC}$-space. Combining this with a result of Section 5 then implies that every generalized Heisenberg group is also a $\mathfrak{TC}$-space. In the final Section 10 we prove that in generalized Heisenberg groups the metric tensor with respect to normal coordinates has the same eigenvalues at antipodal points (with respect to the center of the normal coordinates). This answers the question whether Riemannian g.o. spaces are characterized by this property of the metric tensor in the negative.

3.1 Generalized Heisenberg algebras and groups

3.1.1 Definition

Let $\mathfrak{v}$ and $\mathfrak{z}$ be real vector spaces of dimensions $n, m \in \mathbb{N}$, respectively, and $\beta : \mathfrak{v} \times \mathfrak{v} \longrightarrow \mathfrak{z}$ a skew-symmetric bilinear map. We endow the direct sum $\mathfrak{n} = \mathfrak{v} \oplus \mathfrak{z}$ with an inner product $<.,.>$ such that $\mathfrak{v}$ and $\mathfrak{z}$ are perpendicular and define an $\mathbb{R}$-algebra homomorphism

$$J : \mathfrak{z} \longrightarrow \mathrm{End}(\mathfrak{v}) \ , \ Z \mapsto J_Z$$

by

$$\forall\, U, V \in \mathfrak{v}, Z \in \mathfrak{z} : <J_Z U, V> = <\beta(U,V), Z> .$$

We define a Lie algebra structure on $\mathfrak{n}$ by

$$\forall\, U, V \in \mathfrak{v}, X, Y \in \mathfrak{z} : [U+X, V+Y] := \beta(U,V) .$$

The Lie algebra $\mathfrak{n}$ is said to be a *generalized Heisenberg algebra* if

$$\forall\, Z \in \mathfrak{z} : J_Z^2 = -<Z,Z> id_{\mathfrak{v}} .$$

The attached simply connected Lie group N, endowed with the induced left-invariant Riemannian metric g, is called a *generalized Heisenberg group.*

3.1.2 Classification

Let $J : \mathfrak{z} \longrightarrow \mathrm{End}(\mathfrak{v})$, $Z \mapsto J_Z$ be an $\mathbb{R}$-algebra homomorphism satisfying $J_Z^2 = -<Z,Z> id_{\mathfrak{v}}$ for all $Z \in \mathfrak{z}$. Then J can be extended to an $\mathbb{R}$-algebra homomorphism J^{Cl} from the Clifford algebra $Cl(\mathfrak{z}, q)$ of $\mathfrak{z}$ with respect to the quadratic form $q(Z) := -<Z,Z>$ into $\mathrm{End}(\mathfrak{v})$. Thus J induces a representation J^{Cl} of the Clifford algebra $Cl(\mathfrak{z}, q)$ in $\mathfrak{v}$, that is, $\mathfrak{v}$ is a Clifford module over $Cl(\mathfrak{z}, q)$. Conversely, suppose that J^{Cl} is a representation of a Clifford algebra $Cl(\mathfrak{z}, q)$ in $\mathfrak{v}$, where $\mathfrak{z}$ is an m-dimensional vector space and q is a negative definite quadratic form on $\mathfrak{z}$. The restriction of J^{Cl} to $\mathfrak{z}$ gives an $\mathbb{R}$-algebra homomorphism $J : \mathfrak{z} \longrightarrow \mathrm{End}(\mathfrak{v})$ satisfying $J_Z^2 = q(Z) id_{\mathfrak{v}}$ for all $Z \in \mathfrak{z}$. Equip $\mathfrak{z}$ with the inner product obtained by polarization of $-q$ and extend this to an inner product $<.,.>$ on $\mathfrak{n} := \mathfrak{v} \oplus \mathfrak{z}$ so that $\mathfrak{v}$ and $\mathfrak{z}$ are perpendicular and J_Z acts as an orthogonal map for unit vectors $Z \in \mathfrak{z}$. Then define a skew-symmetric bilinear map $\beta : \mathfrak{v} \times \mathfrak{v} \longrightarrow \mathfrak{z}$ by $<\beta(U,V), Z> := <J_Z U, V>$ for all $U, V \in \mathfrak{v}$ and $Z \in \mathfrak{z}$. In this way we obtain a generalized Heisenberg algebra $\mathfrak{n}$. Thus the classification of generalized Heisenberg algebras and the classification of representations of Clifford algebras of real vector spaces equipped with negative definite quadratic forms are in a one-to-one correspondence up to some equivalence which will be explained below. The latter classification is known (see for example [AtBoSh]) and as follows:

a) If $m \not\equiv 3 (\mathrm{mod}\ 4)$, then there exists (up to equivalence) precisely one irreducible Clifford module $\mathfrak{d}$ over $Cl(\mathfrak{z}, q)$. Every Clifford module $\mathfrak{v}$ over $Cl(\mathfrak{z}, q)$ is isomorphic to the k-fold direct sum of $\mathfrak{d}$, that is,

$$\mathfrak{v} \cong \oplus^k \mathfrak{d}.$$

b) If $m \equiv 3 (\text{mod } 4)$, then there exist (up to equivalence) precisely two non-equivalent irreducible Clifford modules $\mathfrak{d}_1$, $\mathfrak{d}_2$ over $Cl(\mathfrak{z}, q)$. The modules $\mathfrak{d}_1$ and $\mathfrak{d}_2$ have the same dimension and every Clifford module $\mathfrak{v}$ over $Cl(\mathfrak{z}, q)$ is isomorphic to

$$\mathfrak{v} \cong (\oplus^{k_1} \mathfrak{d}_1) \oplus (\oplus^{k_2} \mathfrak{d}_2)$$

for some non-negative integers k_1, k_2.

The dimension n_0 of $\mathfrak{d}$ or $\mathfrak{d}_1$, $\mathfrak{d}_2$ can be taken from the following table:

m	$8p$	$8p+1$	$8p+2$	$8p+3$	$8p+4$	$8p+5$	$8p+6$	$8p+7$
n_0	2^{4p}	2^{4p+1}	2^{4p+2}	2^{4p+2}	2^{4p+3}	2^{4p+3}	2^{4p+3}	2^{4p+3}

Thus we may conclude that for each $m \in \mathbb{N}$ there exist an infinite number of non-isomorphic generalized Heisenberg algebras with $\dim \mathfrak{z} = m$. For the first numbers $m \in \{1, \ldots, 11\}$ we get the following dimensions for N:

m	$\dim N$										
1	3	5	7	9	11	13	15	17	19	21	...
2	6	10	14	18	22	26	30	34	38	42	...
3	7	11	15	19	23	27	31	35	39	43	...
4	12	20	28	36	44	52	60	68	76	84	...
5	13	21	29	37	45	53	61	69	77	85	...
6	14	22	30	38	46	54	62	70	78	86	...
7	15	23	31	39	47	55	63	71	79	87	...
8	24	40	56	72	88	104	120	136	152	168	...
9	41	73	105	137	169	201	233	265	297	329	...
10	74	138	202	266	330	394	458	522	586	650	...
11	75	139	203	267	331	395	459	523	587	651	...

If $m \equiv 3 (\text{mod } 4)$, any two pairs (k_1, k_2) and $(\tilde{k}_1, \tilde{k}_2)$ of non-negative integers with $k_1 + k_2 = \tilde{k}_1 + \tilde{k}_2$ yield generalized Heisenberg algebras $\mathfrak{n}(k_1, k_2)$ and $\mathfrak{n}(\tilde{k}_1, \tilde{k}_2)$ of the same dimension $(k_1 + k_2)n_0 + m$. These are isomorphic if and only if $(\tilde{k}_1, \tilde{k}_2) \in \{(k_1, k_2), (k_2, k_1)\}$. In terms of the groups this means that $N(k_1, k_2)$ and $N(\tilde{k}_1, \tilde{k}_2)$ are isometric if and only if $(\tilde{k}_1, \tilde{k}_2) \in \{(k_1, k_2), (k_2, k_1)\}$. Thus, if for example $m = 3$, there exist two non-isometric eleven-dimensional generalized Heisenberg groups with three-dimensional center.

3.1.3 Algebraic features

Let $\mathfrak{n}$ be a generalized Heisenberg algebra. From the construction follows immediately that $\mathfrak{z}$ is the center of $\mathfrak{n}$. Further, since $[\mathfrak{n}, \mathfrak{n}] = \mathfrak{z}$ and $[\mathfrak{z}, \mathfrak{n}] = 0$ we see that $\mathfrak{n}$ is two-step nilpotent. Thus

Proposition *Every generalized Heisenberg algebra is a two-step nilpotent Lie algebra with center* $\mathfrak{z}$.

It is well-known that every non-singular skew-symmetric bilinear map $\beta : \mathfrak{v} \times \mathfrak{v} \to \mathbb{R}$ has a matrix representation

$$\begin{pmatrix} 0 & I \\ -I & 0 \end{pmatrix}$$

with respect to a suitable basis of $\mathfrak{v}$. So, if the dimension of the center is one, $\mathfrak{n}$ is isomorphic to a Heisenberg algebra in the classical sense. Explicitly, if $n = 2k$ for some $k \in \mathbb{N}$, consider the Lie algebra of all matrices of the form

$$\begin{pmatrix} 0 & x_1 & \dots & x_k & z \\ & & & & y_1 \\ & & 0 & & \vdots \\ & & & & y_k \\ & & & & 0 \end{pmatrix} \quad (x, y \in \mathbb{R}^k,\ z \in \mathbb{R})$$

with Lie bracket given by the usual commutator of matrices. This Lie algebra is the $(n+1)$-dimensional Heisenberg algebra.

From now on U, V, W will always denote vectors in $\mathfrak{v}$ and X, Y, Z vectors in $\mathfrak{z}$. Vectors in $\mathfrak{n}$ will always be written in the form $U + X$. We define

$$|U + X| := \sqrt{<U + X, U + X>}\ .$$

Recall that we have the defining relations

$$J_X^2 = -|X|^2 id_{\mathfrak{v}}$$

and

$$<J_X U, V> = <[U, V], X>\ .$$

Polarization of the first equation gives

$$J_X J_Y + J_Y J_X = -2<X, Y>id_{\mathfrak{v}}\ .$$

Interchanging U and V in the second equation shows that J_X is a skew-symmetric endomorphism, that is,

$$<J_X U, V> + <U, J_X V> = 0\ .$$

Replacing here V by $J_X V$ implies

$$<J_X U, J_X V> = |X|^2 <U, V>\ .$$

Another consequence of the skew-symmetry of J_X and the third equation is

$$<J_X U, J_Y V> + <J_Y U, J_X V> = 2<U, V><X, Y>\ .$$

Replacing here V by U gives

$$<J_X U, J_Y U> = |U|^2 <X, Y>\ .$$

The equation before the last one is equivalent to

$$[J_X U, V] - [U, J_X V] = -2\langle U, V\rangle X \ .$$

Replacing V by $J_X V$ implies

$$[J_X U, J_X V] = -|X|^2[U, V] - 2\langle U, J_X V\rangle X \ ,$$

and replacing U by V gives

$$[V, J_X V] = |V|^2 X \ .$$

These equations will be used frequently in the subsequent sections without referring to them explicitly.

Let $V \in \mathfrak{v}$ be a non-zero vector. We denote by $\ker \mathrm{ad}(V)$ the kernel of the linear map

$$\mathrm{ad}(V) : \mathfrak{v} \to \mathfrak{z} \ , \ U \mapsto [U, V]$$

and by $\ker \mathrm{ad}(V)^\perp$ the orthogonal complement of $\ker \mathrm{ad}(V)$ in $\mathfrak{v}$. Since $U \in \ker \mathrm{ad}(V)$ if and only if $0 = \langle [V, U], Z\rangle = \langle J_Z V, U\rangle$ for all $Z \in \mathfrak{z}$, we see that

$$\ker \mathrm{ad}(V)^\perp = J_{\mathfrak{z}} V \ .$$

If, in addition, V is a unit vector, the map

$$\mathfrak{z} \to \ker \mathrm{ad}(V)^\perp \ , \ Z \mapsto J_Z V$$

is a linear isometry with inverse map

$$\ker \mathrm{ad}(V)^\perp \to \mathfrak{z} \ , \ U \mapsto [V, U] \ .$$

Of particular importance are the generalized Heisenberg algebras which satisfy the so-called *J^2-condition*: For all $X, Y \in \mathfrak{z}$ with $\langle X, Y\rangle = 0$ and all non-zero $U \in \mathfrak{v}$ there exists a $Z \in \mathfrak{z}$ so that $J_X J_Y U = J_Z U$, that is, so that $J_X J_Y U \in \ker \mathrm{ad}(U)^\perp$. Then we have

Theorem [CoDoKoRi] *A generalized Heisenberg algebra $\mathfrak{n}$ satisfies the J^2-condition if and only if*

(i) $m = 1$ *, or*

(ii) $m = 3$ *and* $\mathfrak{n} = \mathfrak{n}(k, 0) \cong \mathfrak{n}(0, k)$ *for some* $k \in \mathbb{N}$*, or*

(iii) $m = 7$ *and* $\mathfrak{n} = \mathfrak{n}(1, 0) \cong \mathfrak{n}(0, 1)$.

These particular generalized Heisenberg algebras are isomorphic to the nilpotent part in the Iwasawa decomposition of the Lie algebra of the isometry group of $\mathbb{C}H^{n+1}$*,* $\mathbb{H}H^{n+1}$ *and* $\mathrm{Cay}H^2$*, respectively.*

3.1.4 Lie exponential map

From Proposition 3.1.3 we immediately get

Proposition 1 *Every generalized Heisenberg group is a two-step nilpotent Lie group.*

Let N be a generalized Heisenberg group. We consider elements in $\mathfrak{n}$ also as left-invariant vector fields on N. For $U + X \in \mathfrak{n}$ denote by γ_{U+X} the one-parameter group generated by $U + X$. Then the Lie exponential map $\exp_{\mathfrak{n}}$ is defined by

$$\exp_{\mathfrak{n}} : \mathfrak{n} \to N \ , \ U + X \mapsto \gamma_{U+X}(1) \ .$$

As the Lie exponential map of every connected, simply connected, nilpotent Lie group is a diffeomorphism (see for example [Hel1, p. 269] or [Rag, p. 6]), we conclude

Proposition 2 *The Lie exponential map* $\exp_{\mathfrak{n}} : \mathfrak{n} \to N$ *is a diffeomorphism.*

This implies

Corollary N *is diffeomorphic to* $\mathbb{R}^{m+n}$.

The group structure on N can be described via $\exp_{\mathfrak{n}}$ by using the Campbell-Hausdorff formula. The two-step nilpotency of N gives the simple formula

$$\exp_{\mathfrak{n}}(U + X) \cdot \exp_{\mathfrak{n}}(V + Y) = \exp_{\mathfrak{n}}\left(U + V + X + Y + \frac{1}{2}[U, V]\right) \ .$$

3.1.5 Some global coordinates

We now introduce some global coordinates on N. Let $V_1, \ldots, V_n, Y_1, \ldots, Y_m$ be an orthonormal basis of the Lie algebra $\mathfrak{n}$ and $\tilde{v}_1, \ldots, \tilde{v}_n, \tilde{y}_1, \ldots, \tilde{y}_m$ the corresponding coordinate functions on $\mathfrak{n}$. The diffeomorphism $\exp_{\mathfrak{n}} : \mathfrak{n} \to N$ then yields global coordinates $v_1, \ldots, v_n, y_1, \ldots, y_m$ on N via the relation

$$(v_1, \ldots, v_n, y_1, \ldots, y_m) \circ \exp_{\mathfrak{n}} = (\tilde{v}_1, \ldots, \tilde{v}_n, \tilde{y}_1, \ldots, \tilde{y}_m) \ .$$

Lemma *We have*

$$\begin{aligned} V_i &= \frac{\partial}{\partial v_i} - \frac{1}{2}\sum_{j,k} A_{ij}^k v_j \frac{\partial}{\partial y_k} \ , \\ Y_i &= \frac{\partial}{\partial y_i} \ , \end{aligned}$$

where

$$A_{ij}^k := \langle [V_i, V_j], Y_k \rangle \ .$$

Proof. Let $p := \exp_{\mathfrak{n}}(U+X) \in N$ be arbitrary and L_p the left translation on N by p. Using the formula for the multiplication in N according to 3.1.4 we obtain

$$\begin{aligned} V_i(p) &= L_{p*e}V_i(e) \\ &= L_{p*e}\left(\frac{\partial}{\partial t}(t \mapsto \exp_{\mathfrak{n}}(tV_i))(0)\right) \\ &= \frac{\partial}{\partial t}(t \mapsto L_p(\exp_{\mathfrak{n}}(tV_i)))(0) \\ &= \dot{\alpha}_i(0) \end{aligned}$$

with

$$\alpha_i(t) = \exp_{\mathfrak{n}}\left(U + tV_i + X + \frac{1}{2}t[U,V_i]\right) .$$

We have

$$\begin{aligned} (v_k \circ \alpha_i)(t) &= \tilde{v}_k(U) + \delta_{ik}t , \\ (y_k \circ \alpha_i)(t) &= \tilde{y}_k(X) + \frac{1}{2}t<[U,V_i],Y_k> \\ &= \tilde{y}_k(X) - \frac{1}{2}t<J_{Y_k}V_i,U> \\ &= \tilde{y}_k(X) - \frac{1}{2}t\sum_j <J_{Y_k}V_i,V_j><U,V_j> \\ &= \tilde{y}_k(X) - \frac{1}{2}t\sum_j A_{ij}^k \tilde{v}_j(U) . \end{aligned}$$

Thus,

$$\begin{aligned} (v_k \circ \alpha_i)'(0) &= \delta_{ik} , \\ (y_k \circ \alpha_i)'(0) &= -\frac{1}{2}\sum_j A_{ij}^k \tilde{v}_j(U) , \end{aligned}$$

and therefore,

$$\begin{aligned} V_i(p) &= \dot{\alpha}_i(0) \\ &= \sum_k (v_k \circ \alpha_i)'(0)\frac{\partial}{\partial v_k}(p) + \sum_k (y_k \circ \alpha_i)'(0)\frac{\partial}{\partial y_k}(p) \\ &= \frac{\partial}{\partial v_i}(p) - \frac{1}{2}\sum_{j,k} A_{ij}^k v_j(p)\frac{\partial}{\partial y_k}(p) . \end{aligned}$$

Similarily,

$$\begin{aligned} Y_i(p) &= L_{p*e}Y_i(e) \\ &= L_{p*e}\left(\frac{\partial}{\partial t}(t \mapsto \exp_{\mathfrak{n}}(tY_i))(0)\right) \\ &= \frac{\partial}{\partial t}(t \mapsto L_p(\exp_{\mathfrak{n}}(tY_i)))(0) \\ &= \dot{\beta}_i(0) \end{aligned}$$

with

$$\beta_i(t) = \exp_{\mathfrak{n}}(U + X + tY_i) .$$

Here we have

$$(v_k \circ \beta_i)(t) = \tilde{v}_k(U) , \; (y_k \circ \beta_i)(t) = \tilde{y}_k(X) + \delta_{ik} t ,$$

and hence

$$(v_k \circ \beta_i)'(0) = 0 , \; (y_k \circ \beta_i)'(0) = \delta_{ik} .$$

This implies

$$Y_i(p) = \dot{\beta}_i(0) = \frac{\partial}{\partial y_i}(p) .$$

Thus the assertion is proved. □

3.1.6 Levi Civita connection

Let ∇ be the Levi Civita connection of a generalized Heisenberg group (N, g). The connection ∇ is completely determined by its values on the left-invariant vector fields in $\mathfrak{n}$. Since g is left-invariant, we may compute ∇ by

$$2g(\nabla_{V+Y}(U + X), W + Z) = g([V, U], Z) - g([U, W], Y) + g([W, V], X)$$

and obtain

$$\nabla_{V+Y}(U + X) = -\frac{1}{2}J_X V - \frac{1}{2}J_Y U - \frac{1}{2}[U, V] .$$

3.1.7 Curvature

Let R be the Riemannian curvature tensor, Q and ρ the Ricci tensor of type $(1,1)$ and $(0,2)$, respectively, and τ the scalar curvature of a generalized Heisenberg group (N, g). All these objects are completely determined by its values on the tangent space $T_e N \cong \mathfrak{n}$ of N at the identity e. By a straightforward computation one gets

$$\begin{aligned}
&R(U + X, V + Y)(W + Z) \\
&\qquad = -\frac{1}{4}J_{[V,W]}U + \frac{1}{4}J_{[U,W]}V + \frac{1}{2}J_{[U,V]}W \\
&\qquad\quad -\frac{1}{4}J_Y J_Z U + \frac{1}{4}J_X J_Z V + \frac{1}{2}J_X J_Y W + \frac{1}{2}\langle X, Y\rangle W \\
&\qquad\quad +\frac{1}{4}[V, J_X W] - \frac{1}{4}[U, J_Y W] - \frac{1}{2}[U, J_Z V] + \frac{1}{2}\langle U, V\rangle Z, \\
Q(U + X) &= -\frac{m}{2}U + \frac{n}{4}X , \\
(\nabla_{V+Y}Q)(U + X) &= -\frac{2m + n}{8}(J_X V - [U, V]) , \\
\tau &= -\frac{1}{4}mn .
\end{aligned}$$

Further, if $U+X, V+Y \in \mathfrak{n}$ are orthonormal and σ is the span of $U+X$ and $V+Y$, then the sectional curvature $K(\sigma)$ of N with respect to σ is

$$K(\sigma) = \frac{1}{4}(|U|^2|Y|^2 + |V|^2|X|^2) + <X,Y><U,V> - \frac{3}{4}|[U,V]|^2 - \frac{3}{2}<J_XU, J_YV> .$$

In particular, it can easily be seen that the sectional curvature attains both positive and negative values.

The equation for ∇Q shows that $(\nabla_{V+Y}\rho)(V+Y, V+Y) = 0$, that is, the Ricci tensor is invariant under the geodesic flow of N. Polarization of the latter equation shows that the cyclic sum over all entries in $\nabla\rho$ vanishes, that is, the Ricci tensor of N is a Killing tensor.

Lemma *The Ricci tensor of any generalized Heisenberg group is a Killing tensor, or equivalently, is invariant under the geodesic flow.*

As ∇Q never vanishes, we also have

Proposition *None of the generalized Heisenberg groups is a locally symmetric space.*

3.1.8 The Jacobi operator

One of the central objects in our studies is the Jacobi operator defined by

$$R_{V+Y} := R(.,V+Y)(V+Y)$$

for all $V+Y \in \mathfrak{n}$. The above expression for the curvature tensor implies as a special case

$$\begin{aligned} R_{V+Y}(U+X) &= \frac{3}{4}J_{[U,V]}V + \frac{3}{4}J_XJ_YV + \frac{1}{4}|Y|^2U + \frac{1}{2}<X,Y>V \\ &\quad -\frac{3}{4}[U,J_YV] + \frac{1}{4}|V|^2X + \frac{1}{2}<U,V>Y . \end{aligned}$$

By a straightforward computation we get for the covariant derivative

$$R'_{V+Y} := (\nabla_{V+Y}R)(.,V+Y)(V+Y)$$

of the Jacobi operator R_{V+Y} the expression

$$\begin{aligned} R'_{V+Y}(U+X) &= \frac{3}{2}J_{[U,J_YV]}V + \frac{3}{2}J_{[U,V]}J_YV - \left(\frac{1}{2}|V|^2 + |Y|^2\right)J_XV \\ &\quad -(<U,V> - <X,Y>)J_YV - <U,J_YV>V \\ &\quad + \left(\frac{1}{2}|V|^2 + |Y|^2\right)[U,V] + <U,J_YV>Y . \end{aligned}$$

The spectrum of the Jacobi operator will be computed explicitly in 3.3.

3.1.9 Geodesics

Let $V + Y \in T_e N \cong \mathfrak{n}$ be a unit vector and $\gamma : \mathbb{R} \to N$ the geodesic in N with $\gamma(0) = e$ and $\dot\gamma(0) = V + Y$. We consider subspaces of $\mathfrak{n}$ also as subbundles of TN via left translation.

Since $\nabla_V V = 0$ and $\nabla_Y Y = 0$, the integral curves of V and Y are geodesics in N. This implies that

$$\gamma(t) = \exp_{\mathfrak{n}}(tV) \quad , \text{ if } Y = 0 ,$$
$$\gamma(t) = \exp_{\mathfrak{n}}(tY) \quad , \text{ if } V = 0 .$$

Now suppose that $V \neq 0 \neq Y$. The vectors V, $J_Y V$ and Y span a three-dimensional Heisenberg algebra $\mathfrak{n}_3$. It is easy to check that $\mathfrak{n}_3$ is an autoparallel subbundle of TN. Therefore γ lies in the leaf of $\mathfrak{n}_3$ through e, which is a totally geodesically embedded three-dimensional Heisenberg group. This shows

Proposition *Every geodesic γ in N lies in a totally geodesically embedded three-dimensional Heisenberg group. The latter one is uniquely determined if and only if $\dot\gamma$ is not tangent to $\mathfrak{v}$ or $\mathfrak{z}$.*

So, the determination of geodesics in generalized Heisenberg groups can be reduced to the one in a three-dimensional Heisenberg group.

We continue with the case $V \neq 0 \neq Y$ and put

$$\hat V := \frac{V}{|V|} \ , \ \hat Y := \frac{Y}{|Y|} \ .$$

Then γ is of the form

$$\gamma(t) = \exp_{\mathfrak{n}}(a(t)\hat V + b(t) J_{\hat Y} \hat V + c(t)\hat Y)$$

with some functions a, b, c satisfying

$$a(0) = 0 \ , \ a'(0) = |V| \ , \ b(0) = 0 \ , \ b'(0) = 0 \ , \ c(0) = 0 \ , \ c'(0) = |Y| \ .$$

Let u, v, x be the global coordinates on the three-dimensional Heisenberg group determined by $\hat V, J_{\hat Y}\hat V, \hat Y$ and in accordance with 3.1.5. By means of Lemma 3.1.5 we have

$$\begin{aligned} \frac{\partial}{\partial u} &= \hat V + \frac{1}{2} v \hat Y \ , \\ \frac{\partial}{\partial v} &= J_{\hat Y}\hat V - \frac{1}{2} u \hat Y \ , \\ \frac{\partial}{\partial x} &= \hat Y \ , \end{aligned}$$

and therefore,

$$\begin{aligned} \dot\gamma &= (u \circ \gamma)' \frac{\partial}{\partial u} \circ \gamma + (v \circ \gamma)' \frac{\partial}{\partial v} \circ \gamma + (x \circ \gamma)' \frac{\partial}{\partial x} \circ \gamma \\ &= a'\hat V + b' J_{\hat Y}\hat V + \left(\frac{1}{2}(a'b - ab') + c'\right) \hat Y \ . \end{aligned}$$

Differentiating this again gives the system of equations

$$\begin{aligned}
0 &= a'' + \frac{1}{2}a'bb' - \frac{1}{2}ab'^2 + b'c' \,, \\
0 &= b'' - \frac{1}{2}a'^2 b + \frac{1}{2}aa'b' - a'c' \,, \\
0 &= \frac{1}{2}(a'b - ab')' + c'' \,.
\end{aligned}$$

The last equation implies

$$c' = \frac{1}{2}(ab' - a'b) + |Y| \,.$$

Inserting this into the first two ones then gives

$$a'' = -|Y|b' \,,\; b'' = |Y|a'$$

and thus

$$a'' = -|Y|^2 a \,.$$

Successively we may now compute $a(t)$, $b(t)$, $c(t)$, and we obtain

$$\begin{aligned}
a(t) &= \frac{|V|}{|Y|}\sin(|Y|t) \,, \\
b(t) &= \frac{|V|}{|Y|}(1 - \cos(|Y|t)) \,, \\
c(t) &= |Y|t + \frac{|V|^2}{2|Y|}\left(t - \frac{1}{|Y|}\sin(|Y|t)\right) \,.
\end{aligned}$$

Summing up, we obtain (see also [Kap2] and the correction in [Kap3])

Theorem *Let $V + Y \in T_e N \cong \mathfrak{n}$ be a unit vector and $\gamma : \mathbb{R} \to N$ the geodesic in N with $\gamma(0) = e$ and $\dot\gamma(0) = V + Y$. Then*

$$\begin{aligned}
\gamma(t) &= \exp_{\mathfrak{n}}(tV) \,,\; \textit{if } Y = 0 \,, \\
\gamma(t) &= \exp_{\mathfrak{n}}(tY) \,,\; \textit{if } V = 0 \,, \\
\gamma(t) &= \exp_{\mathfrak{n}}\left(\frac{1}{|Y|}\sin(|Y|t)V + \frac{1}{|Y|^2}(1 - \cos(|Y|t))J_Y V \right. \\
&\quad \left. + \left(t + \frac{|V|^2}{2|Y|^2}\left(t - \frac{1}{|Y|}\sin(|Y|t)\right)\right)Y\right) \,,\; \textit{if } V \neq 0 \neq Y \,.
\end{aligned}$$

In addition, when we identify different tangent spaces of N along γ via left translation, we have

$$\begin{aligned}
\dot\gamma(t) &= V \,,\; \textit{if } Y = 0 \,, \\
\dot\gamma(t) &= Y \,,\; \textit{if } V = 0 \,, \\
\dot\gamma(t) &= \cos(|Y|t)V + \frac{1}{|Y|}\sin(|Y|t)J_Y V + Y \,,\; \textit{if } V \neq 0 \neq Y \,.
\end{aligned}$$

3.1.10 Integrability of $\mathfrak{v}$ and $\mathfrak{z}$

Consider $\mathfrak{v}$ and $\mathfrak{z}$ as left-invariant distributions of TN.

Proposition *The distribution $\mathfrak{v}$ is not integrable. The distribution $\mathfrak{z}$ is autoparallel and hence integrable. Each leaf of $\mathfrak{z}$ is a totally geodesically embedded $\mathbb{R}^m$ ($m = \dim \mathfrak{z}$) endowed with the standard Euclidean metric.*

Proof. The non-integrability of $\mathfrak{v}$ follows from $[\mathfrak{v},\mathfrak{v}] = \mathfrak{z}$. The explicit expression of the Levi Civita connection shows that

$$\forall\, X,Y \in \mathfrak{z} : \nabla_Y X = 0 \ ,$$

whence $\mathfrak{z}$ is autoparallel. Therefore $\mathfrak{z}$ is integrable and each leaf M of $\mathfrak{z}$ is a totally geodesic submanifold of N. As

$$\forall\, X,Y,Z \in \mathfrak{z} : R(X,Y)Z = 0 \ ,$$

the Gauss equation of second order implies that M is flat. Recall that the Lie exponential map $\exp_{\mathfrak{n}} : \mathfrak{n} \to N$ is a diffeomorphism. For each $Y \in \mathfrak{z}$ the integral curve through e is a geodesic in N. Thus, the leaf of $\mathfrak{z}$ through e is $\exp_{\mathfrak{n}}(\mathfrak{z})$, which is diffeomorphic to $\mathbb{R}^m$. Eventually, by left-invariance, we see that each leaf of $\mathfrak{z}$ is diffeomorphic to $\mathbb{R}^m$. □

3.1.11 Irreducibility

Proposition *Every generalized Heisenberg group is irreducible as a Riemannian manifold.*

Proof. Suppose that a generalized Heisenberg group (N,g) is a Riemannian product $M_1 \times M_2$ with $\dim M_i \geq 1$. Let $V_1 + Y_1$ and $V_2 + Y_2$ be tangent vectors to M_1 and M_2 at some point $p \in N$, respectively, where the decomposition is with respect to the decomposition of T_pN into $\mathfrak{v} \oplus \mathfrak{z}$ obtained by left translation from e. Then

$$0 = (\nabla_{V_1+Y_1}\rho)(V_1+Y_1, V_2+Y_2) = -\frac{2m+n}{8}g(J_{Y_1}V_1, V_2) = \frac{2m+n}{8}g(J_{Y_1}V_2, V_1) \ .$$

This shows that $J_{Y_1}V_1$ is tangent to M_1 and $J_{Y_1}V_2$ is tangent to M_2 at p. Analogously, $J_{Y_2}V_2$ is tangent to M_2 and $J_{Y_2}V_1$ is tangent to M_1 at p. As $0 = g(V_1,V_2) + g(Y_1,Y_2)$, this implies

$$\begin{aligned} 0 &= g(J_{Y_1}V_1, J_{Y_2}V_2) + g(J_{Y_2}V_1, J_{Y_1}V_2) \\ &= 2g(Y_1,Y_2)g(V_1,V_2) \ = \ -2g(Y_1,Y_2)^2 \ = \ -2g(V_1,V_2)^2 \ . \end{aligned}$$

Therefore V_1, Y_1 are tangent to M_1 and V_2, Y_2 are tangent to M_2 at p. Now it is clear that there exist $i,j \in \{1,2\}$, $i \neq j$, such that V_i and Y_j may be chosen as unit vectors. The plane σ spanned by V_i and Y_j has sectional curvature equal to zero, but, on the other hand, the formula for $K(\sigma)$ in 3.1.7 gives $K(\sigma) = 1/4$, which is a contradiction. □

3.1.12 The operator K

Let $V+Y$ be a vector in $\mathfrak{n}$ with $V \neq 0 \neq Y$ and let $Y^\perp$ denote the orthogonal complement of the span of Y in $\mathfrak{z}$. We put

$$\hat{V} := \frac{V}{|V|}\,,\ \hat{Y} := \frac{Y}{|Y|}\,,$$

and define an endomorphism

$$K_{V,Y} : Y^\perp \to Y^\perp, X \mapsto [\hat{V}, J_X J_{\hat{Y}} \hat{V}]\,.$$

For $X, Z \in Y^\perp$ we have

$$\begin{aligned} <K_{V,Y}X, Z> &= <[\hat{V}, J_X J_{\hat{Y}}\hat{V}], Z> = <J_Z\hat{V}, J_X J_{\hat{Y}}\hat{V}> = -<J_X\hat{V}, J_Z J_{\hat{Y}}\hat{V}> \\ &= -<[\hat{V}, J_Z J_{\hat{Y}}\hat{V}], X> = -<K_{V,Y}Z, X>\,, \end{aligned}$$

which shows that $K_{V,Y}$ is skew-symmetric. Thus $K_{V,Y}^2$ is a symmetric endomorphism of $Y^\perp$. Let X be an eigenvector of $K_{V,Y}^2$ of unit length with corresponding eigenvalue μ. Then

$$\mu = <K_{V,Y}^2 X, X> = -<K_{V,Y}X, K_{V,Y}X> = -\|[\hat{V}, J_X J_{\hat{Y}}\hat{V}]\|^2\,.$$

Putting

$$J_X J_{\hat{Y}} \hat{V} = U + J_Z \hat{V}$$

with $U \in \ker \mathrm{ad}(\hat{V})$ and $Z \in \mathfrak{z}$ then gives

$$\mu = -\|[\hat{V}, J_Z\hat{V}]\|^2 = -|Z|^2\,.$$

As

$$1 = |J_X J_{\hat{Y}}\hat{V}|^2 = |U|^2 + |Z|^2\,,$$

we deduce that $\mu \in [-1, 0]$ and $\sqrt{-\mu}$ is the length of the projection of $J_X J_{\hat{Y}}\hat{V}$ onto $\ker \mathrm{ad}(\hat{V})^\perp = \ker \mathrm{ad}(V)^\perp$. In particular,

$$\begin{aligned} \mu = 0 &\iff J_X J_Y V \in \ker \mathrm{ad}(V)\,, \\ \mu = -1 &\iff J_X J_Y V \in \ker \mathrm{ad}(V)^\perp\,. \end{aligned}$$

In the latter case we then have

$$J_X J_Y V = |Y| J_{K_{V,Y}X} V\,.$$

Further, we see that $\mathfrak{n}$ satisfies the J^2-condition if and only if $K_{V,Y}^2 = -id_{Y^\perp}$ for all $V + Y \in \mathfrak{n}$ with $V \neq 0 \neq Y$.

Suppose now that $V + Y$ is a unit vector and let $\gamma : \mathbb{R} \to N$ be the geodesic in N with $\gamma(0) = e$ and $\dot{\gamma}(0) = V + Y$. Then

$$\dot{\gamma}(t) = V(t) + Y$$

with

$$V(t) = \cos(|Y|t)V + \frac{1}{|Y|}\sin(|Y|t)J_Y V \ ,$$

where different tangent spaces along γ are identified via left translation. We define a skew-symmetric $Y^\perp$-valued tensor field K_γ along γ by

$$K_\gamma(t) := K_{V(t),Y} \ .$$

Then

$$\begin{aligned}
K_\gamma(t)X &= [\hat V(t), J_X J_{\hat Y}\hat V(t)] \\
&= \cos^2(|Y|t)[\hat V, J_X J_{\hat Y}\hat V] + \sin^2(|Y|t)[J_{\hat Y}\hat V, J_X J_{\hat Y}^2 \hat V] \\
&\quad + \sin(|Y|t)\cos(|Y|t)([\hat V, J_X J_{\hat Y}^2\hat V] + [J_{\hat Y}\hat V, J_X J_{\hat Y}\hat V]) \\
&= [\hat V, J_X J_{\hat Y}\hat V] \\
&= K_\gamma(0)X \ ,
\end{aligned}$$

since

$$[J_{\hat Y}\hat V, J_X J_{\hat Y}^2\hat V] = -[J_{\hat Y}\hat V, J_X\hat V] = -[J_X J_{\hat Y}\hat V, \hat V]$$

and

$$[\hat V, J_X J_{\hat Y}^2\hat V] + [J_{\hat Y}\hat V, J_X J_{\hat Y}\hat V] = -[\hat V, J_X\hat V] + X = 0 \ .$$

This implies

Lemma *The eigenvalues of K_γ^2 are constant along γ.*

3.1.13 Isometry group

We denote by $A(N)$ the group of automorphisms of N whose differential at e is an orthogonal map and by $L(N)$ the group consisting of the left translations on N. Both $A(N)$ and $L(N)$ act on N via isometries. Then we have

Proposition [Kap2] *The isometry group of a generalized Heisenberg group (N, g) is the semidirect product of $A(N)$ and $L(N)$, where $A(N)$ acts on $L(N)$ via conjugation.*

The algebraic structure of $A(N)$ has been determined by C. Riehm in [Rie1].

3.1.14 Kähler structures

The question whether there exists a Kähler structure on some generalized Heisenberg group has to be answered in the negative. In fact, according to Lemma 3.1.7 the Ricci tensor of a generalized Heisenberg group N is a Killing tensor. If N carries the structure of a Kähler manifold compatible with the left-invariant Riemannian metric g, then its Ricci tensor must necessarily be parallel (see [SeVa3]). Since none of the generalized Heisenberg groups has a parallel Ricci tensor, we conclude:

Proposition *None of the generalized Heisenberg groups carries a Kähler structure which is compatible with its left-invariant Riemannian metric.*

3.2 Some classifications

In this section we present the known classifications of generalized Heisenberg groups which belong to one of the various classes of symmetric-like Riemannian manifolds discussed in Chapter 2.

Theorem 1 [Kap3] *A generalized Heisenberg group is a naturally reductive Riemannian homogeneous space if and only if* $\dim \mathfrak{z} \in \{1,3\}$.

See also [TrVal] for an alternative proof.

Theorem 2 [Rie2] *A generalized Heisenberg group N is a Riemannian g.o. space if and only if*

(i) $\dim \mathfrak{z} \in \{1,2,3\}$, *or*

(ii) $\dim \mathfrak{z} = 5$ *and* $\dim N = 13$, *or*

(iii) $\dim \mathfrak{z} = 6$ *and* $\dim N = 14$, *or*

(iv) $\dim \mathfrak{z} = 7$ *and*

- *(1)* $\dim N = 15$, *or*
- *(2)* $\dim N = 23$ *and* $\mathfrak{n} = \mathfrak{n}(2,0) \cong \mathfrak{n}(0,2)$, *or*
- *(3)* $\dim N = 31$ *and* $\mathfrak{n} = \mathfrak{n}(3,0) \cong \mathfrak{n}(0,3)$.

For $\dim \mathfrak{z} = 2$ and $\dim N = 6$ see also [TrVal] and [Kap3].

F. Ricci [Ric] has classified all generalized Heisenberg groups N for which the convolution algebra $L^1_{A(N)}$ is commutative; here, $A(N)$ is the group of automorphisms of N that act as orthogonal transformations on $\mathfrak{n}$ and $L^1_{A(N)}$ is the algebra of all L^1-functions that are invariant under $A(N)$. According to A. Kaplan and F. Ricci [KaRi], this convolution algebra is commutative if and only if the algebra of all invariant differential operators is commutative. Hence we get

Theorem 3 [Ric] *A generalized Heisenberg group N is a commutative space if and only if*

(i) $\dim \mathfrak{z} \in \{1,2,3\}$, *or*

(ii) $\dim \mathfrak{z} = 5$ *and* $\dim N = 13$, *or*

(iii) $\dim \mathfrak{z} = 6$ *and* $\dim N = 14$, *or*

(iv) $\dim \mathfrak{z} = 7$ *and*

- *(1)* $\dim N = 15$, *or*
- *(2)* $\dim N = 23$ *and* $\mathfrak{n} = \mathfrak{n}(2,0) \cong \mathfrak{n}(0,2)$.

From the preceding two theorems we get an example of a Riemannian g.o. space which is not a commutative space, namely the 31-dimensional generalized Heisenberg group N with seven-dimensional center and $\mathfrak{n} \cong \mathfrak{n}(3,0)$.

In 4.1.10 it will be shown that N is isometric to a horosphere in a complex hyperbolic space if $\dim \mathfrak{z} = 1$, in a quaternionic hyperbolic space if $\dim \mathfrak{z} = 3$ and $\mathfrak{n} = \mathfrak{n}(0,k) \cong \mathfrak{n}(k,0)$ for some $k \in \mathbb{N}$, and in Cayley hyperbolic plane if $\dim \mathfrak{z} = 7$ and $\dim N = 15$. So from Theorem 2 in 2.3 we obtain

Proposition 1 *A generalized Heisenberg group N is a weakly symmetric space if*

(i) $\dim \mathfrak{z} = 1$, *or*

(ii) $\dim \mathfrak{z} = 3$ *and* $\mathfrak{n} = \mathfrak{n}(0,k) \cong \mathfrak{n}(k,0)$ *for some* $k \in \mathbb{N}$, *or*

(iii) $\dim \mathfrak{z} = 7$ *and* $\dim N = 15$.

As a weakly symmetric space is always commutative (see 2.4), one could therefore check the remaining cases in Theorem 3 in order to obtain a complete classification of the weakly symmetric generalized Heisenberg groups. Up to now we were not able to settle this question.

Finally we state

Theorem 4 [Kap3] *Every generalized Heisenberg group is a D'Atri space.*

In Sections 6, 9 and 10 we will give alternative proofs of this result.

As every harmonic space is an Einstein manifold, but none of the generalized Heisenberg groups is an Einstein space, we obtain

Proposition 2 *None of the generalized Heisenberg groups is a harmonic space.*

Up to now we do not know which of the generalized Heisenberg groups are probabilistic commutative. We also mention that in [TrVa1] and [Kap3] the geometry of the six-dimensional generalized Heisenberg group with two-dimensional center is studied and further properties are obtained. In particular, it is proved that the eigenvalues of (g_{ij}) in normal coordinates have antipodal symmetry. In Section 10 we will show that this property holds in fact on every generalized Heisenberg group.

3.3 Spectral properties of the Jacobi operator

In this section we compute the eigenvalues and the corresponding eigenspaces of the Jacobi operators of an arbitrary generalized Heisenberg group N at the identity e.

Theorem *Let $V + Y$ be a unit vector in $\mathfrak{n}$.*

(i) $\underline{Y = 0}$. *Then R_V has three distinct eigenvalues* 0, $-3/4$ *and* $1/4$; *the corresponding eigenspaces are* $\ker \operatorname{ad}(V)$, $\ker \operatorname{ad}(V)^{\perp}$ *and* $\mathfrak{z}$, *respectively.*

(ii) $\underline{V = 0}$. *Then* R_Y *has two distinct eigenvalues* 0 *and* 1/4*; the corresponding eigenspaces are* $\mathfrak{z}$ *and* $\mathfrak{v}$*, respectively.*

(iii) $\underline{V \neq 0 \neq Y}$. *We decompose* $\mathfrak{n}$ *orthogonally into*

$$\mathfrak{n} = \mathfrak{n}_3 \oplus \mathfrak{p} \oplus \mathfrak{q} \ ,$$

where

$$\begin{aligned} \mathfrak{n}_3 &:= \operatorname{span}\{V, J_Y V, Y\} \ , \\ \mathfrak{p} &:= \ker \operatorname{ad}(V) \cap \ker \operatorname{ad}(J_Y V) \ , \\ \mathfrak{q} &:= \operatorname{span}\{Y^\perp, J_{Y^\perp} V, J_{Y^\perp} J_Y V\} \ . \end{aligned}$$

The spaces $\mathfrak{n}_3$*,* $\mathfrak{p}$ *and* $\mathfrak{q}$ *are invariant under the action of* R_{V+Y} *and we have:*

(1) $R_{V+Y}|\mathfrak{n}_3$ *has two (if* $|V|^2 = 1/4$*) or three (if* $|V|^2 \neq 1/4$*) distinct eigenvalues, namely*

$$\begin{aligned} 0 \ \textit{and} \ \frac{1}{4} \quad &, \quad \textit{if} \ |V|^2 = \frac{1}{4} \ , \\ 0, \frac{1}{4} \ \textit{and} \ \frac{1}{4} - |V|^2 \quad &, \quad \textit{if} \ |V|^2 \neq \frac{1}{4} \ ; \end{aligned}$$

the corresponding eigenspaces are

$$\begin{aligned} \operatorname{span}\{V + Y, J_Y V\} \ \textit{and} \ \mathbb{R}(-|Y|^2 V + |V|^2 Y) \quad &, \quad \textit{if} \ |Y|^2 = \frac{1}{4} \ , \\ \mathbb{R}(V + Y), \mathbb{R}(-|Y|^2 V + |V|^2 Y) \ \textit{and} \ \mathbb{R} J_Y V \quad &, \quad \textit{if} \ |Y|^2 \neq \frac{1}{4} \ , \end{aligned}$$

respectively.

(2) (if $\mathfrak{p} \neq \{0\}$*)* $R_{V+Y}|\mathfrak{p}$ *has only one eigenvalue, namely* $(1 - |V|^2)/4 = |Y|^2/4$.

(3) (if $\mathfrak{q} \neq \{0\}$*) We put* $K := K_{V,Y}$ *and decompose* $Y^\perp$ *orthogonally into*

$$Y^\perp = L_0 \oplus \ldots \oplus L_k \ ,$$

where

$$L_j := \ker(K^2 - \mu_j id_{Y^\perp}) \ , \quad (j = 0, \ldots, k)$$

and

$$0 \geq \mu_0 > \mu_1 > \ldots > \mu_k \geq -1$$

are the distinct eigenvalues of K^2*. It can easily be seen that*

$$X \in L_j \Longrightarrow KX \in L_j \quad (j = 0, \ldots, k) \ ,$$

whence $\dim L_j$ *is even provided that* $\mu_j \neq 0$*. We now define*

$$\begin{aligned} \mathfrak{q}_j &:= \operatorname{span}\{L_j, J_{L_j} V, J_{L_j} J_Y V\} \ , \ j = 0, \ldots, k, \ \mu_k \neq -1 \ , \\ \mathfrak{q}_k &:= \operatorname{span}\{L_k, J_{L_k} V\} \ , \ \textit{if} \ \mu_k = -1 \ . \end{aligned}$$

Then

$$\mathfrak{q} = \mathfrak{q}_0 \oplus \ldots \oplus \mathfrak{q}_k \ , \ \dim \mathfrak{q}_j \equiv \begin{cases} 0 (\mathrm{mod}\, 3) & , \ if\ \mu_j = 0 \\ 0 (\mathrm{mod}\, 4) & , \ if\ \mu_j = -1 \\ 0 (\mathrm{mod}\, 6) & , \ otherwise \end{cases}$$

and each space $\mathfrak{q}_j$ is invariant under the action of R_{V+Y}. Finally, we put

$$\begin{aligned} \rho_1 &:= \frac{1}{4} - |V|^2 \ , \\ \rho_2 &:= \frac{1}{8}(1 + \sqrt{1 + 32|V|^2|Y|^2}) \ , \\ \rho_3 &:= \frac{1}{8}(1 - \sqrt{1 + 32|V|^2|Y|^2}) \ . \end{aligned}$$

(A) (if $j = k$ and $\mu_k = -1$) $R_{V+Y}|\mathfrak{q}_k$ has two distinct eigenvalues κ_{k1} and κ_{k2}, which are the solutions of

$$\left(\rho - \frac{1}{4}|V|^2\right)(\rho - \rho_1) = \frac{9}{16}|V|^2|Y|^2 \ ;$$

the corresponding eigenspaces are

$$\left\{(\kappa_{ki} - \rho_1)X + \frac{3}{4}|Y|J_{KX}V \mid X \in L_k\right\} \quad (i = 1, 2).$$

We always have

$$\kappa_{k1} + \kappa_{k2} = \frac{1}{4}|V|^2 + \rho_1 = \frac{1}{4}(1 - 3|V|^2) \ .$$

(B) (otherwise) $R_{V+Y}|\mathfrak{q}_j$ has three distinct eigenvalues κ_{j1}, κ_{j2} and κ_{j3}, which are the solutions of

$$(\rho - \rho_1)(\rho - \rho_2)(\rho - \rho_3) = \frac{27}{64}|V|^4|Y|^2\mu_j \ .$$

We always have

$$\kappa_{k1} + \kappa_{k2} + \kappa_{k3} = \rho_1 + \rho_2 + \rho_3 = \frac{1}{2} - |V|^2 \ .$$

The corresponding eigenspaces are given by
(a) (if $j = 0$ and $\mu_0 = 0$)

$$\begin{aligned} &J_{L_0}V && for\ \kappa_{01} := \rho_1 \ , \\ &\{(4\rho_i - |Y|^2)X + 3J_X J_Y V \mid X \in L_0\} && for\ \kappa_{0i} := \rho_i \ (i = 2, 3); \end{aligned}$$

(b) (otherwise)

$$\left\{(\rho_1 - \kappa_{ji})((4\kappa_{ji} - |Y|^2)X + 3J_X J_Y V) + \frac{9}{4}|V|^2|Y|J_{KX}V \mid X \in L_j\right\}.$$

Remark. As $\dim Y^\perp = m-1$, the dimension of $\mathfrak{q}$ can be estimated by

$$2(m-1) \le \dim \mathfrak{q} \le 3(m-1) ,$$

and the first inequality is an equality for all $V+Y \in \mathfrak{n}$ with $V \ne 0 \ne Y$ precisely if $\mathfrak{n}$ satisfies the J^2-condition.

Proof. We recall from 3.1.8 that

$$\begin{aligned} R_{V+Y}(U+X) &= \frac{3}{4}J_{[U,V]}V + \frac{3}{4}J_X J_Y V + \frac{1}{4}|Y|^2 U + \frac{1}{2}\langle X,Y\rangle V \\ &\quad -\frac{3}{4}[U, J_Y V] + \frac{1}{4}|V|^2 X + \frac{1}{2}\langle U,V\rangle Y , \end{aligned}$$

and consider three cases.

(i) $\underline{Y=0}$. Then $|V|^2 = 1$ and

$$R_V(U+X) = \frac{3}{4}J_{[U,V]}V + \frac{1}{4}X .$$

This implies that

$$\forall\, U \in \ker \operatorname{ad}(V) : R_V U = 0$$

and

$$\forall\, X \in \mathfrak{z} : R_V X = \frac{1}{4}X .$$

If $U \in \ker \operatorname{ad}(V)^\perp$, then there exists a $Z \in \mathfrak{z}$ so that $U = J_Z V$ and we get

$$R_V U = \frac{3}{4}J_{[J_Z V,V]}V = -\frac{3}{4}J_Z V = -\frac{3}{4}U .$$

(ii) $\underline{V=0}$. Then $|Y|^2 = 1$ and

$$R_Y(U+X) = \frac{1}{4}U .$$

Thus

$$\forall\, U \in \mathfrak{v} : R_Y U = \frac{1}{4}U$$

and

$$\forall\, X \in \mathfrak{z} : R_Y X = 0 .$$

(iii) $\underline{V \ne 0 \ne Y}$. Since

$$\begin{aligned} J_{Y^\perp} V &= \ker \operatorname{ad}(V)^\perp \cap (J_Y V)^\perp , \\ J_{Y^\perp} J_Y V &= \ker \operatorname{ad}(J_Y V)^\perp \cap (V)^\perp , \end{aligned}$$

it follows easily that

$$\mathfrak{q} = \operatorname{span}\{Y^\perp, J_{Y^\perp}V, J_{Y^\perp}J_Y V\}$$

is orthogonal to $\mathfrak{n}_3$ and $\mathfrak{p}$.

The cases (1) and (2) may be checked without difficulties by a straightforward computation.

From now on we suppose that $\mathfrak{q} \neq \{0\}$. For $X \in L_j$ we have

$$R_{V+Y}X = \begin{cases} \frac{1}{4}|V|^2X + \frac{3}{4}|Y|J_{KX}V & , \text{ if } j = k \text{ and } \mu_k = -1 \\ \frac{1}{4}|V|^2X + \frac{3}{4}J_XJ_YV & , \text{ otherwise ,} \end{cases}$$

$$R_{V+Y}J_XV = \left(\frac{1}{4} - |V|^2\right)J_XV - \frac{3}{4}|V|^2|Y|KX \,,$$

and if $j < k$ or ($j = k$ and $\mu_k \neq -1$),

$$R_{V+Y}J_XJ_YV = \frac{3}{4}|V|^2|Y|^2X - \frac{3}{4}|V|^2|Y|J_{KX}V + \frac{1}{4}|Y|^2J_XJ_YV \,.$$

This shows that $\mathfrak{q}_j$ is invariant under the action of R_{V+Y} for each $j \in \{0,\ldots,k\}$.

Next, we compute the eigenvalues and the eigenspaces of $R_{V+Y}|\mathfrak{q}_j$. We start with the case $j = k$ and $\mu_k = -1$. Let $X \in L_k$ be a non-zero vector. Then we see from the above formulae that span$\{X, J_{KX}V\}$ is invariant under R_{V+Y}. Thus, there exist $\alpha, \beta \in \mathbb{R}$ such that $\alpha X + \beta J_{KX}V$ is an eigenvector of R_{V+Y}, say with corresponding eigenvalue κ. Then we get the equations

$$\begin{aligned} \kappa\alpha &= \frac{1}{4}|V|^2\alpha + \frac{3}{4}|V|^2|Y|\beta \,, \\ \kappa\beta &= \frac{3}{4}|Y|\alpha + \left(\frac{1}{4} - |V|^2\right)\beta = \frac{3}{4}|Y|\alpha + \rho_1\beta \,. \end{aligned}$$

The second equation yields

$$(\kappa - \rho_1)\beta = \frac{3}{4}|Y|\alpha \,.$$

As $\beta = 0$ is impossible (since it implies also $\alpha = 0$), we may normalize the eigenvector such that $\beta = 3|Y|/4$. Then $\alpha = \kappa - \rho_1$, and from the first equation we get

$$\left(\kappa - \frac{1}{4}|V|^2\right)(\kappa - \rho_1) = \frac{9}{16}|V|^2|Y|^2 \,.$$

The assertions stated in (A) now follow easily.

From now on we assume that $j < k$ or $\mu_k \neq -1$. We first consider the case when $j = 0$ and $\mu_0 = 0$. Let X be a non-zero vector in L_0. Then $KX = 0$, as $|KX|^2 = -\langle K^2X, X\rangle = 0$, and hence

$$R_{V+Y}J_XV = \left(\frac{1}{4} - |V|^2\right)J_XV = \rho_1 J_XV \,,$$

whence ρ_1 is an eigenvalue of R_{V+Y} with $J_{L_0}V$ consisting of corresponding eigenvectors. Next, we see that span$\{X, J_XJ_YV\}$ is invariant under R_{V+Y}. Hence, there exist $\alpha, \beta \in \mathbb{R}$ such that $\alpha X + \beta J_X J_Y V$ is an eigenvector of R_{V+Y}. The resulting equations we get in this case are

$$\begin{aligned} \kappa\alpha &= \frac{1}{4}|V|^2\alpha + \frac{3}{4}|V|^2|Y|^2\beta \,, \\ \kappa\beta &= \frac{3}{4}\alpha + \frac{1}{4}|Y|^2\beta \,. \end{aligned}$$

The second equation gives

$$\frac{3}{4}\alpha = \left(\kappa - \frac{1}{4}|Y|^2\right)\beta \,.$$

As $\beta = 0$ is impossible (since it implies $\alpha = 0$), we may normalize the eigenvector so that $\beta = 3/4$. Then $\alpha = \kappa - |Y|^2/4$ and hence

$$\left(\kappa - \frac{1}{4}|V|^2\right)\left(\kappa - \frac{1}{4}|Y|^2\right) = \frac{9}{16}|V|^2|Y|^2 \,,$$

or equivalently,

$$(\kappa - \rho_2)(\kappa - \rho_3) = 0 \,.$$

From this the assertion in (B)(a) follows immediately.

It remains to study the case $j > 0$ or $\mu_0 \neq 0$. Let X be a non-zero vector in L_j. From the above computations we see that there exist $\alpha, \beta, \gamma \in \mathbb{R}$ such that

$$\alpha X + \beta J_{KX} V + \gamma J_X J_Y V$$

is an eigenvector of R_{V+Y}. Here, the eigenvector equations are

$$\begin{aligned} \kappa\alpha &= \frac{1}{4}|V|^2\alpha - \frac{3}{4}|V|^2|Y|\mu_j\beta + \frac{3}{4}|V|^2|Y|^2\gamma \,, \\ \kappa\beta &= \rho_1\beta - \frac{3}{4}|V|^2|Y|\gamma \,, \\ \kappa\gamma &= \frac{3}{4}\alpha + \frac{1}{4}|Y|^2\gamma \,. \end{aligned}$$

The second equation gives

$$(\kappa - \rho_1)\beta = -\frac{3}{4}|V|^2|Y|\gamma \,.$$

As $\beta = 0$ is impossible (since this implies $\gamma = 0$ and hence $\alpha = 0$), we may normalize the eigenvector by putting $\beta = 9|V|^2|Y|/4$. Then

$$\gamma = 3(\rho_1 - \kappa)$$

and hence, using the third equation,

$$\alpha = (4\kappa - |Y|^2)(\rho_1 - \kappa) \,.$$

The first equation then gives

$$\left(\kappa - \frac{1}{4}|V|^2\right)(4\kappa - |Y|^2)(\rho_1 - \kappa) = \frac{9}{4}|V|^2|Y|^2(\rho_1 - \kappa) - \frac{27}{16}|V|^4|Y|^2\mu_j \,,$$

that is,

$$\begin{aligned}\frac{27}{64}|V|^4|Y|^2\mu_j &= \left(\frac{9}{16}|V|^2|Y|^2 - \left(\kappa - \frac{1}{4}|V|^2\right)\left(\kappa - \frac{1}{4}|Y|^2\right)\right)(\rho_1 - \kappa) \\ &= (\kappa - \rho_1)(\kappa - \rho_2)(\kappa - \rho_3) .\end{aligned}$$

From this the assertion in (B)(b) then follows. □

An immediate consequence of the Theorem is

Corollary *None of the generalized Heisenberg groups is a pointwise Osserman space.*

3.4 Constancy of the spectrum along geodesics

We shall now prove that in any generalized Heisenberg group the eigenvalues of the Jacobi operator are constant along geodesics.

Theorem *Every generalized Heisenberg group is a $\mathfrak{C}$-space.*

We will provide two alternative proofs, one using eigenvalues and the other one using eigenvectors of the Jacobi operator along geodesics.

First proof. Let $V + Y \in \mathfrak{n}$ be a unit vector and $\gamma : \mathbb{R} \to N$ the geodesic in N with $\gamma(0) = e$ and $\dot\gamma(0) = V + Y$. According to Theorem 3.1.9 we have

$$\dot\gamma(t) = \begin{cases} V(t) + Y & , \text{ if } Y \neq 0 \\ V & , \text{ if } Y = 0 , \end{cases}$$

where

$$V(t) = \cos(|Y|t)V + \frac{1}{|Y|}\sin(|Y|t)J_Y V .$$

Thus $|V(t)|$ and $|Y(t)|$, the lengths of the projections of $\dot\gamma(t)$ onto $\mathfrak{v}$ and $\mathfrak{z}$, respectively, are constant. Further, from Lemma 3.1.12, we know that the eigenvalues of K_γ^2 are constant along γ. As according to Theorem 3.3 the eigenvalues of the Jacobi operator R_γ along γ depend only on $|V(t)|$, $|Y(t)|$ and the eigenvalues of K_γ^2, we conclude that the eigenvalues of R_γ are constant. □

Second proof. We shall apply characterization (iv) in Proposition 2 of 2.8. Let $V + Y \in \mathfrak{n}$ be a unit vector. We have to prove that for each eigenvalue κ of R_{V+Y} there exists a corresponding eigenvector $U + X$ so that $<R'_{V+Y}(U+X), U+X> = 0$. In order to accomplish this we use the expressions for the eigenspaces of R_{V+Y} which have been computed in Theorem 3.3. From 3.1.8 we recall that

$$\begin{aligned} R'_{V+Y}(U+X) &= \frac{3}{2}J_{[U,J_Y V]}V + \frac{3}{2}J_{[U,V]}J_Y V - \left(\frac{1}{2}|V|^2 + |Y|^2\right)J_X V \\ &\quad -(<U,V> - <X,Y>)J_Y V - <U,J_Y V>V \\ &\quad + \left(\frac{1}{2}|V|^2 + |Y|^2\right)[U,V] + <U,J_Y V>Y \end{aligned}$$

and consider again three cases.

(i) $\underline{Y=0}$. Then

$$R'_V(U+X) = -\frac{1}{2}J_XV + \frac{1}{2}[U,V]$$

and hence,

$$\forall\, U \in \ker \mathrm{ad}(V) : \langle R'_V U, U\rangle = 0 \,,$$
$$\forall\, U \in \ker \mathrm{ad}(V)^\perp : \langle R'_V U, U\rangle = \frac{1}{2}\langle [U,V], U\rangle = 0 \,,$$
$$\forall\, X \in \mathfrak{z} : \langle R'_V X, X\rangle = -\frac{1}{2}\langle J_X V, X\rangle = 0 \,.$$

(ii) $\underline{V=0}$. In this case the assertion follows from $R'_Y = 0$.

(iii) $\underline{V \neq 0 \neq Y}$. On $\mathfrak{n}_3$ we have

$$\begin{aligned} R'_{V+Y}(V+Y) &= 0 \,, \\ R'_{V+Y}J_YV &= -\frac{1}{2}|V|^2(-|Y|^2V + |V|^2Y) \,, \\ R'_{V+Y}(-|Y|^2V + |V|^2Y) &= -\frac{1}{2}|V|^2 J_Y V \,. \end{aligned}$$

From this we conclude that $\langle R'_{V+Y}(U+X), U+X\rangle = 0$ for $U+X$ being one of the eigenvectors $V+Y$, J_YV and $-|Y|^2V + |V|^2Y$.

On $\mathfrak{p}$ the assertion holds obviously since $R'_{V+Y}|\mathfrak{p} = 0$.

Finally, to verify the assertion on $\mathfrak{q}$, let X be a unit vector in some L_j. Then

$$\begin{aligned} R'_{V+Y}X &= -\left(\frac{1}{2}|V|^2 + |Y|^2\right) J_X V \,, \\ R'_{V+Y}J_XV &= \frac{3}{2}|V|^2|Y|J_{KX}V - \frac{3}{2}|V|^2 J_X J_Y V - |V|^2\left(\frac{1}{2}|V|^2 + |Y|^2\right) X \,, \\ R'_{V+Y}J_{KX}V &= \frac{3}{2}|V|^2|Y|\mu_j J_X V - \frac{3}{2}|V|^2 J_{KX} J_Y V - |V|^2\left(\frac{1}{2}|V|^2 + |Y|^2\right) KX \,, \\ R'_{V+Y}J_XJ_YV &= -\frac{3}{2}|V|^2|Y|^2 J_X V - \frac{3}{2}|V|^2|Y| J_{KX}J_YV \\ &\quad -|V|^2|Y|\left(\frac{1}{2}|V|^2 + |Y|^2\right) KX \,. \end{aligned}$$

First suppose that $j = k$ and $\mu_k = -1$. From the preceding formulae we see that

$$R'_{V+Y}(\mathrm{span}\{X, J_{KX}V\}) \subset \mathrm{span}\{KX, J_XV\} \perp \mathrm{span}\{X, J_{KX}V\} \,,$$

from which the assertion follows.

Next, if $j = 0$ and $\mu_0 = 0$, then

$$R'_{V+Y}J_XV \perp J_XV$$

and

$$R'_{V+Y}(\mathrm{span}\{X, J_X J_Y V\}) \subset \mathrm{span}\{J_X V\} \perp \mathrm{span}\{X, J_X J_Y V\} ,$$

which implies again the assertion.

Finally, in the remaining cases we have

$$\begin{aligned} R'_{V+Y}(\mathrm{span}\{X, J_{KX} V, J_X J_Y V\}) &\subset \mathrm{span}\{KX, J_X V, J_{KX} J_Y V\} \\ &\perp \mathrm{span}\{X, J_{KX} V, J_X J_Y V\} , \end{aligned}$$

which also implies the required result. □

According to Proposition 3.1.7, none of the generalized Heisenberg groups is a locally symmetric space. As a $\mathfrak{C}$-space, which is in addition also a $\mathfrak{P}$-space, is necessarily locally symmetric (see 2.8), we get

Corollary *None of the generalized Heisenberg groups is a $\mathfrak{P}$-space.*

3.5 Rotation of the eigenspaces along geodesics

In this section we concentrate on the behavior of the eigenspaces of the Jacobi operator along geodesics in generalized Heisenberg groups. The tensor field T_γ constructed in this section will be of fundamental importance for the computation of Jacobi fields in 3.7. Let $V + Y \in \mathfrak{n}$ be a unit vector and

$$\mathfrak{q} := \mathrm{span}\{Y^\perp, J_{Y^\perp} V, J_{Y^\perp} J_Y V\} .$$

Thus

$$\mathfrak{q} = \begin{cases} \ker \mathrm{ad}(V)^\perp \oplus \mathfrak{z} & , \text{ if } Y = 0 \\ Y^\perp & , \text{ if } V = 0 , \end{cases}$$

and $\mathfrak{q}$ is as in Theorem 3.3 if $V \neq 0 \neq Y$. Further,

$$\mathfrak{q}^\perp = \begin{cases} \ker \mathrm{ad}(V) & , \text{ if } Y = 0 \\ \mathfrak{v} \oplus \mathbb{R}Y & , \text{ if } V = 0 \\ \mathfrak{n}_3 \oplus \mathfrak{p} & , \text{ if } V \neq 0 \neq Y \end{cases}$$

with $\mathfrak{n}_3$ and $\mathfrak{p}$ as in Theorem 3.3. Next, let $\gamma : \mathbb{R} \to N$ be the geodesic in N with $\gamma(0) = e$ and $\dot\gamma(0) = V + Y$. Then

$$\dot\gamma(t) = \begin{cases} V(t) + Y & , \text{ if } Y \neq 0 \\ V & , \text{ if } Y = 0 \end{cases}$$

with

$$V(t) = \cos(|Y|t)V + \frac{1}{|Y|}\sin(|Y|t) J_Y V .$$

In the following computations we will use the equation

$$\frac{d}{dt}(t \mapsto V(t)) = J_Y V(t) .$$

If not stated otherwise, different tangent spaces along γ will be identified by left translation.

We now define a tensor field T_γ along γ by

$$T_\gamma(t)(U+X) = \begin{cases} \frac{3}{2}J_Y U + \frac{1}{2}J_X V(t) + \frac{1}{2}[U, V(t)] & , \text{ if } U+X \in \mathfrak{q} \\ -\frac{1}{2}J_Y U + \frac{1}{2}J_X V(t) + \frac{1}{2}[U, V(t)] & , \text{ if } U+X \in \mathfrak{q}^\perp . \end{cases}$$

Lemma 1 *T_γ is a skew-symmetric parallel tensor field along γ.*

Proof. If $Y = 0$, then

$$T_\gamma(t)(U+X) = \frac{1}{2}J_X V + \frac{1}{2}[U, V] .$$

T_γ is clearly skew-symmetric and

$$\begin{aligned} T'_\gamma(t)(U+X) &= \nabla_V(T_\gamma(t)(U+X)) - T_\gamma(t)\nabla_V(U+X) \\ &= \frac{1}{2}\nabla_V J_X V + \frac{1}{2}\nabla_V[U,V] + \frac{1}{2}T_\gamma(t)J_X V + \frac{1}{2}T_\gamma(t)[U,V] \\ &= -\frac{1}{4}[J_X V, V] - \frac{1}{4}J_{[U,V]}V + \frac{1}{4}[J_X V, V] + \frac{1}{4}J_{[U,V]}V \\ &= 0 . \end{aligned}$$

If $V = 0$, then

$$T_\gamma(t)(U+X) = -\frac{1}{2}J_Y U .$$

Also here, T_γ is skew-symmetric and

$$\begin{aligned} T'_\gamma(t)(U+X) &= \nabla_Y(T_\gamma(t)(U+X)) - T_\gamma(t)\nabla_Y(U+X) \\ &= -\frac{1}{2}\nabla_Y J_Y U + \frac{1}{2}T_\gamma(t)J_Y U \\ &= -\frac{1}{4}U + \frac{1}{4}U = 0 . \end{aligned}$$

Now, let $V \neq 0 \neq Y$. If $U + X \in \mathfrak{q}^\perp$, then

$$X = \frac{1}{|Y|^2}\langle X, Y\rangle Y \quad \text{and} \quad U \in (\ker \mathrm{ad}(V) \cap \ker \mathrm{ad}(J_Y V)) \oplus \mathbb{R}V \oplus \mathbb{R}J_Y V .$$

This implies that

$$J_Y U \ , \ J_X V(t) \ , \ [U, V(t)] \in \mathfrak{q}^\perp ,$$

and hence

$$T_\gamma(t)(U+X) = -\frac{1}{2}J_Y U + \frac{1}{2}J_X V(t) + \frac{1}{2}[U, V(t)] \in \mathfrak{q}^\perp .$$

Thus, T_γ maps $\mathfrak{q}^\perp$ into itself. It can easily be checked that $T_\gamma|\mathfrak{q}^\perp$ is skew-symmetric. Further, we have

$$\begin{aligned}
T'_\gamma(t)(U+X) &= \nabla_{V(t)+Y}(T_\gamma(t)(U+X)) - T_\gamma(t)\nabla_{V(t)+Y}(U+X) \\
&= -\frac{1}{2}\nabla_{V(t)+Y} J_Y U + \frac{1}{2}\nabla_{V(t)+Y} J_X V(t) + \frac{1}{2}\nabla_{V(t)+Y}[U, V(t)] \\
&\quad +\frac{1}{2}T_\gamma(t) J_Y U + \frac{1}{2}T_\gamma(t) J_X V(t) + \frac{1}{2}T_\gamma(t)[U, V(t)] \\
&= \frac{1}{4}J_Y^2 U + \frac{1}{4}[J_Y U, V(t)] + \frac{1}{2}J_X J_Y V(t) - \frac{1}{4}J_Y J_X V(t) \\
&\quad -\frac{1}{4}[J_X V(t), V(t)] + \frac{1}{2}[U, J_Y V(t)] - \frac{1}{4}J_{[U,V(t)]}V(t) \\
&\quad -\frac{1}{4}J_Y^2 U + \frac{1}{4}[J_Y U, V(t)] - \frac{1}{4}J_Y J_X V(t) \\
&\quad +\frac{1}{4}[J_X V(t), V(t)] + \frac{1}{4}J_{[U,V(t)]}V(t) \\
&= \frac{1}{2}J_X J_Y V(t) - \frac{1}{2}J_Y J_X V(t) + \frac{1}{2}[U, J_Y V(t)] + \frac{1}{2}[J_Y U, V(t)] \\
&= 0\,,
\end{aligned}$$

since

$$J_X J_Y V(t) - J_Y J_X V(t) = \frac{1}{|Y|^2}<X, Y>(J_Y^2 V(t) - J_Y^2 V(t)) = 0$$

and, as $U \in (\ker \mathrm{ad}(V) \cap \ker \mathrm{ad}(J_Y V)) \oplus \mathbb{R}V \oplus \mathbb{R}J_Y V$,

$$\begin{aligned}
\frac{1}{2}([J_Y U, V(t)] + [U, J_Y V(t)]) &= [J_Y U, V(t)] + <U, V(t)>Y \\
&= \frac{1}{|Y|^2}<[J_Y U, V(t)], Y>Y + <U, V(t)>Y \\
&= \frac{1}{|Y|^2}<J_Y^2 U, V(t)>Y + <U, V(t)>Y = 0\,.
\end{aligned}$$

Next, if $U + X \in \mathfrak{q}$, then also

$$J_Y U \;,\; J_X V(t)\;,\; [U, V(t)] \in \mathfrak{q}\,,$$

and hence

$$T_\gamma(t)(U+X) = \frac{3}{2}J_Y U + \frac{1}{2}J_X V(t) + \frac{1}{2}[U, V(t)] \in \mathfrak{q}\,.$$

So, T_γ maps $\mathfrak{q}$ into $\mathfrak{q}$. Again, it follows easily that $T_\gamma|\mathfrak{q}$ is skew-symmetric. Finally, we obtain

$$\begin{aligned}
T'_\gamma(t)(U+X) &= \nabla_{V(t)+Y} T_\gamma(t)(U+X) - T_\gamma(t)\nabla_{V(t)+Y}(U+X) \\
&= \frac{3}{2}\nabla_{V(t)+Y} J_Y U + \frac{1}{2}\nabla_{V(t)+Y} J_X V(t) + \frac{1}{2}\nabla_{V(t)+Y}[U, V(t)] \\
&\quad +\frac{1}{2}T_\gamma(t) J_Y U + \frac{1}{2}T_\gamma(t) J_X V(t) + \frac{1}{2}T_\gamma(t)[U, V(t)]
\end{aligned}$$

$$
\begin{aligned}
&= -\frac{3}{4}J_Y^2U - \frac{3}{4}[J_YU, V(t)] + \frac{1}{2}J_XJ_YV(t) - \frac{1}{4}J_YJ_XV(t) \\
&\quad -\frac{1}{4}[J_XV(t), V(t)] + \frac{1}{2}[U, J_YV(t)] - \frac{1}{4}J_{[U,V(t)]}V(t) \\
&\quad +\frac{3}{4}J_Y^2U + \frac{1}{4}[J_YU, V(t)] + \frac{3}{4}J_YJ_XV(t) \\
&\quad +\frac{1}{4}[J_XV(t), V(t)] + \frac{1}{4}J_{[U,V(t)]}V(t) \\
&= \frac{1}{2}J_XJ_YV(t) + \frac{1}{2}J_YJ_XV(t) + \frac{1}{2}[U, J_YV(t)] - \frac{1}{2}[J_YU, V(t)] \\
&= 0\,,
\end{aligned}
$$

for $<X, Y> = 0$ and $<U, V(t)> = 0$, and this finishes the proof. □

Lemma 2 $R'_\gamma = [R_\gamma, T_\gamma]$.

Proof. If $Y = 0$, then

$$
\begin{aligned}
R_\gamma(t)(U + X) &= R_V(U + X) = \frac{3}{4}J_{[U,V]}V + \frac{1}{4}X\,, \\
R'_\gamma(t)(U + X) &= R'_V(U + X) = -\frac{1}{2}J_XV + \frac{1}{2}[U, V]\,, \\
T_\gamma(t)(U + X) &= \frac{1}{2}J_XV + \frac{1}{2}[U, V]\,.
\end{aligned}
$$

Thus

$$
\begin{aligned}
[R_\gamma, T_\gamma](t)(U + X) &= \frac{3}{8}J_{[J_XV,V]}V + \frac{1}{8}[U, V] - \frac{1}{8}J_XV - \frac{3}{8}[J_{[U,V]}V, V] \\
&= -\frac{1}{2}J_XV + \frac{1}{2}[U, V] \\
&= R'_\gamma(t)(U + X)\,.
\end{aligned}
$$

If $V = 0$, we have

$$
\begin{aligned}
R_\gamma(t)(U + X) &= R_Y(U + X) = \frac{1}{4}U\,, \\
R'_\gamma(t)(U + X) &= R'_Y(U + X) = 0\,, \\
T_\gamma(U + X) &= -\frac{1}{2}J_YU\,.
\end{aligned}
$$

Therefore,

$$
[R_\gamma, T_\gamma](t)(U + X) = -\frac{1}{8}J_YU + \frac{1}{8}J_YU = 0 = R'_\gamma(t)(U + X)\,.
$$

Next, let $V \neq 0 \neq Y$. First suppose that $U + X \in \mathfrak{q}$. Then

$$
R_\gamma(t)(U + X) = \frac{3}{4}J_{[U,V(t)]}V(t) + \frac{3}{4}J_XJ_YV(t) + \frac{1}{4}|Y|^2U - \frac{3}{4}[U, J_YV(t)] + \frac{1}{4}|V|^2X\,,
$$

and each term on the right-hand side is in $\mathfrak{q}$. Further,

$$\begin{aligned} R'_\gamma(t)(U+X) &= \frac{3}{2}J_{[U,J_YV(t)]}V(t) + \frac{3}{2}J_{[U,V(t)]}J_YV(t) - \left(\frac{1}{2}|V|^2 + |Y|^2\right)J_XV(t) \\ &\quad + \left(\frac{1}{2}|V|^2 + |Y|^2\right)[U,V(t)] \end{aligned}$$

and

$$T_\gamma(t)(U+X) = \frac{3}{2}J_YU + \frac{1}{2}J_XV(t) + \frac{1}{2}[U,V(t)] \ .$$

We now obtain

$$\begin{aligned} [R_\gamma, T_\gamma](t)(U+X) &= \frac{9}{8}J_{[J_YU,V(t)]}V(t) + \frac{3}{8}|Y|^2J_YU - \frac{9}{8}[J_YU, J_YV(t)] \\ &\quad + \frac{3}{8}J_{[J_XV(t),V(t)]}V(t) + \frac{1}{8}|Y|^2J_XV(t) - \frac{3}{8}[J_XV(t), J_YV(t)] \\ &\quad + \frac{3}{8}J_{[U,V(t)]}J_YV(t) + \frac{1}{8}|V|^2[U,V(t)] \\ &\quad - \frac{9}{8}J_YJ_{[U,V(t)]}V(t) - \frac{3}{8}[J_{[U,V(t)]}V(t), V(t)] \\ &\quad - \frac{9}{8}J_YJ_XJ_YV(t) - \frac{3}{8}[J_XJ_YV(t), V(t)] \\ &\quad - \frac{3}{8}|Y|^2J_YU - \frac{1}{8}|Y|^2[U,V(t)] \\ &\quad + \frac{3}{8}J_{[U,J_YV(t)]}V(t) - \frac{1}{8}|V|^2J_XV(t) \\ &= R'_\gamma(t)(U+X) \ , \end{aligned}$$

since

$$\begin{aligned} &\frac{9}{8}J_{[J_YU,V(t)]}V(t) + \frac{3}{8}J_{[U,J_YV(t)]}V(t) = \frac{3}{2}J_{[U,J_YV(t)]}V(t) \ , \\ &\frac{3}{8}J_{[U,V(t)]}J_YV(t) - \frac{9}{8}J_YJ_{[U,V(t)]}V(t) = \frac{3}{2}J_{[U,V(t)]}J_YV(t) \ , \\ &\frac{3}{8}J_{[J_XV(t),V(t)]}V(t) - \frac{1}{8}|V|^2J_XV(t) = -\frac{1}{2}|V|^2J_XV(t) \ , \\ &\frac{1}{8}|Y|^2J_XV(t) - \frac{9}{8}J_YJ_XJ_YV(t) = -|Y|^2J_XV(t) \ , \\ &\frac{1}{8}|V|^2[U,V(t)] - \frac{3}{8}[J_{[U,V(t)]}V(t), V(t)] = \frac{1}{2}|V|^2[U,V(t)] \ , \\ &-\frac{9}{8}[J_YU, J_YV(t)] - \frac{1}{8}|Y|^2[U,V(t)] = |Y|^2[U,V(t)] \ , \\ &-\frac{3}{8}[J_XV(t), J_YV(t)] - \frac{3}{8}[J_XJ_YV(t), V(t)] = 0 \ . \end{aligned}$$

Finally, let $U + X \in \mathfrak{q}^\perp$, that is,

$$X = \frac{1}{|Y|^2}\langle X, Y\rangle Y \quad \text{and} \quad U \in (\ker \operatorname{ad}(V) \cap \ker \operatorname{ad}(J_YV)) \oplus \mathbb{R}V \oplus \mathbb{R}J_YV \ .$$

Then

$$[U,V(t)] = \frac{1}{|Y|^2}<[U,V(t)],Y>Y = \frac{1}{|Y|^2}<J_Y U, V(t)>Y = -\frac{1}{|Y|^2}<U, J_Y V(t)>Y$$

and

$$[U, J_Y V(t)] = \frac{1}{|Y|^2}<[U, J_Y V(t)], Y>Y = \frac{1}{|Y|^2}<J_Y U, J_Y V(t)>Y = <U, V(t)>Y .$$

Therefore,

$$\begin{aligned} R_\gamma(t)(U+X) &= \frac{3}{4}J_{[U,V(t)]}V(t) + \frac{3}{4}J_X J_Y V(t) + \frac{1}{4}|Y|^2 U + \frac{1}{2}<X,Y>V(t) \\ &\quad -\frac{3}{4}[U, J_Y V(t)] + \frac{1}{4}|V|^2 X + \frac{1}{2}<U,V(t)>Y \\ &= -\frac{1}{4}<X,Y>V(t) - \frac{3}{4|Y|^2}<U, J_Y V(t)> J_Y V(t) + \frac{1}{4}|Y|^2 U \\ &\quad -\frac{1}{4}<U,V(t)>Y + \frac{1}{4}|V|^2 X , \end{aligned}$$

and each term on the right-hand side is in $\mathfrak{q}^\perp$. Further,

$$\begin{aligned} R'_\gamma(t)(U+X) &= \frac{3}{2}J_{[U,J_Y V(t)]}V(t) + \frac{3}{2}J_{[U,V(t)]}J_Y V(t) - \left(\frac{1}{2}|V|^2 + |Y|^2\right) J_X V(t) \\ &\quad -(<U,V(t)> - <X,Y>)J_Y V(t) - <U, J_Y V(t)>V(t) \\ &\quad + \left(\frac{1}{2}|V|^2 + |Y|^2\right)[U,V(t)] + <U, J_Y V(t)>Y \\ &= \frac{1}{2}<U, J_Y V(t)>V(t) + \frac{1}{2}<U,V(t)>J_Y V(t) - \frac{1}{2}|V|^2 J_X V(t) \\ &\quad -\frac{|V|^2}{2|Y|^2}<U, J_Y V(t)>Y \end{aligned}$$

and

$$\begin{aligned} T_\gamma(t)(U+X) &= -\frac{1}{2}J_Y U + \frac{1}{2}J_X V(t) + \frac{1}{2}[U,V(t)] \\ &= -\frac{1}{2}J_Y U + \frac{1}{2}J_X V(t) - \frac{1}{2|Y|^2}<U, J_Y V(t)>Y . \end{aligned}$$

This yields

$$\begin{aligned} &[R_\gamma, T_\gamma](t)(U+X) \\ &= \frac{3}{8}<U,V(t)>J_Y V(t) - \frac{1}{8}|Y|^2 J_Y U - \frac{1}{8}<U, J_Y V(t)>Y - \frac{3}{8}|V|^2 J_X V(t) \\ &\quad +\frac{1}{8}|Y|^2 J_X V(t) + \frac{1}{8}<U, J_Y V(t)>V(t) - \frac{|V|^2}{8|Y|^2}<U, J_Y V(t)>Y \\ &\quad -\frac{1}{8}|Y|^2 J_X V(t) + \frac{3}{8}<U, J_Y V(t)>V(t) - \frac{3|V|^2}{8|Y|^2}<U, J_Y V(t)>Y + \frac{1}{8}|Y|^2 J_Y U \\ &\quad +\frac{1}{8}<U, J_Y V(t)>Y + \frac{1}{8}<U,V(t)>J_Y V(t) - \frac{1}{8}|V|^2 J_X V(t) \end{aligned}$$

$$
\begin{aligned}
&= \frac{1}{2}<U, J_Y V(t)>V(t) + \frac{1}{2}<U, V(t)>J_Y V(t) - \frac{1}{2}|V|^2 J_X V(t) \\
&\quad - \frac{|V|^2}{2|Y|^2}<U, J_Y V(t)>Y \\
&= R'_\gamma(t)(U + X) ,
\end{aligned}
$$

and this completes the proof. □

The combination of Lemmas 1 and 2 now provides

Theorem *Every generalized Heisenberg group is a $\mathfrak{C}_0$-space.*

Proposition 1 in 2.9 shows that

$$R_\gamma(t) = e^{-tT_\gamma(0)} \circ R_\gamma(0) \circ e^{tT_\gamma(0)} ,$$

where different tangent spaces along γ are identified via parallel translation. Thus the rotation of the eigenspaces of R_γ is described by the one-parameter subgroup $e^{tT_\gamma(0)}$ of the orthogonal group $O(\mathfrak{n})$. An analogous statement holds also for the higher order Jacobi operators $R_\gamma^{(k)}$ (see 2.9).

3.6 Some corollaries

We will now draw some conclusions from the results in the preceding section. First of all, according to Theorem 2 in 3.2, not every generalized Heisenberg group is a Riemannian g.o. space. On the other hand, according to Proposition 2 in 2.9, every Riemannian g.o. space is a $\mathfrak{C}_0$-space. So Theorem 3.5 implies

Corollary 1 *There exist $\mathfrak{C}_0$-spaces which are not Riemannian g.o. spaces.*

Next, from Proposition 5 in 2.11 and Theorem 3.5 we may conclude:

Corollary 2 *A generalized Heisenberg group is a $\mathfrak{TC}$-space if and only if it is an $\mathfrak{SC}$-space.*

This corollary will be useful in Section 3.9.

As any $\mathfrak{C}_0$-space is also a $\mathfrak{C}$-space, Theorem 3.5 implies an alternative proof of

Corollary 3 *Every generalized Heisenberg group is a $\mathfrak{C}$-space.*

According to Theorem 4 in 3.2, every generalized Heisenberg group is a D'Atri space. The proof of this, provided by A. Kaplan, uses harmonic analysis. From Proposition 3 in 2.9 we know that every $\mathfrak{C}_0$-space is a D'Atri space. Thus we now have an alternative proof of

Corollary 4 *Every generalized Heisenberg group is a D'Atri space.*

One of the open problems concerning $\mathfrak{C}$-spaces is the question whether they are locally homogeneous or not. An obvious way to tackle this problem is to combine all the operators T_v in the condition $R'_v = [R_v, T_v]$, which characterizes $\mathfrak{C}$-spaces, and to see whether they provide a homogeneous structure (see 2.1). Now, consider the endomorphism T_{V+Y} defined in 3.5. They always have the property $T_{V+Y}(V+Y) = 0$. If all the operators T_{V+Y} would fit together to provide a homogeneous structure, it would by means of Proposition 2 in 2.1 necessarily be a naturally reductive one. But according to Theorem 1 in 3.2 this is possible only if the dimension of the center is one or three. Thus, we conclude that the endomorphisms T_{V+Y} do in general not fit together to provide a homogeneous structure on a generalized Heisenberg group.

3.7 Jacobi fields

In order to compute the Jacobi fields on generalized Heisenberg groups, we use the following general idea.

Lemma *Let M be a Riemannian manifold with Levi Civita connection ∇ and $\gamma : I \to M$ a geodesic in M parametrized by arc length. We denote by ∂ the standard unit tangent field on I. Suppose there exists a ∇_∂-parallel skew-symmetric tensor field T_γ along γ such that the Jacobi operator R_γ along γ satisfies $\nabla_\partial R_\gamma := R'_\gamma = [R_\gamma, T_\gamma]$. Then define a new covariant derivative*

$$\bar{\nabla}_\partial := \nabla_\partial + T_\gamma \ ,$$

and put

$$\bar{R}_\gamma := R_\gamma + T_\gamma^2 \ .$$

Then R_γ, $\bar{R}_\gamma$ and T_γ are $\bar{\nabla}_\partial$-parallel along γ and the Jacobi equation along γ is

$$\bar{\nabla}_\partial \bar{\nabla}_\partial B - 2T_\gamma \bar{\nabla}_\partial B + \bar{R}_\gamma B = 0 \ .$$

Proof. For any vector field B along γ we have

$$\begin{aligned}
(\bar{\nabla}_\partial R_\gamma)B &= \bar{\nabla}_\partial R_\gamma B - R_\gamma \bar{\nabla}_\partial B \\
&= \nabla_\partial R_\gamma B + T_\gamma R_\gamma B - R_\gamma \nabla_\partial B - R_\gamma T_\gamma B \\
&= (\nabla_\partial R_\gamma - [R_\gamma, T_\gamma])B \\
&= 0 \ , \\
(\bar{\nabla}_\partial T_\gamma)B &= \bar{\nabla}_\partial T_\gamma B - T_\gamma \bar{\nabla}_\partial B \\
&= \nabla_\partial T_\gamma B + T_\gamma^2 B - T_\gamma \nabla_\partial B - T_\gamma^2 B \\
&= (\nabla_\partial T_\gamma)B \\
&= 0 \ .
\end{aligned}$$

Hence R_γ and T_γ are $\bar{\nabla}_\partial$-parallel. This implies that also $\bar{R}_\gamma = R_\gamma + T_\gamma^2$ is $\bar{\nabla}_\partial$-parallel. Using the fact that T_γ is $\bar{\nabla}_\partial$-parallel we now compute for any Jacobi field B along γ:

$$\begin{aligned}
0 &= \nabla_\partial \nabla_\partial B + R_\gamma B \\
&= \bar{\nabla}_\partial \bar{\nabla}_\partial B - T_\gamma \bar{\nabla}_\partial B - \bar{\nabla}_\partial T_\gamma B + T_\gamma^2 B + R_\gamma B \\
&= \bar{\nabla}_\partial \bar{\nabla}_\partial B - 2T_\gamma \bar{\nabla}_\partial B + \bar{R}_\gamma B \ .
\end{aligned}$$

This proves the assertion. □

This lemma is of interest as soon as one wishes to compute explicitly Jacobi fields in non-symmetric Riemannian manifolds. The advantage of introducing $\bar{\nabla}_\theta$ is that with respect to this covariant derivative the Jacobi equation becomes a second order linear equation with *constant* coefficients. Of course, in general, a $\bar{\nabla}_\theta$-parallel tensor field T_γ as in the lemma does not exist. But, as follows from the very definition, on any $\mathfrak{C}_0$-space (and hence on any generalized Heisenberg group) the Jacobi equation can be transformed in such a way. The methods of solving the Jacobi equation on normal homogeneous Riemannian manifolds used by I. Chavel [Cha] or, more generally, on naturally reductive Riemannian homogeneous spaces by W. Ziller [Zil1], who use the canonical connection on these spaces, are a special case of the above procedure.

We shall now apply this method to compute explicitly some particular Jacobi fields on generalized Heisenberg groups.

Theorem *Let $V + Y$ be a unit vector in $\mathfrak{n}$ and $\gamma : \mathbb{R} \to N$ the geodesic in N with $\gamma(0) = e$ and $\dot\gamma(0) = V + Y$. For $U + X \in \mathfrak{n}$ we denote by B_{U+X} the Jacobi field along γ with initial values $B_{U+X}(0) = 0$ and $B'_{U+X}(0) = U + X$. We have*

(i) $\underline{Y = 0}$. *Then*

$$\begin{aligned} B_U(t) &= tU + \frac{1}{2}t^2[U,V] , \\ B_X(t) &= \left(t - \frac{1}{6}t^3\right) X + \frac{1}{2}t^2 J_X V \end{aligned}$$

for all $U \in \mathfrak{v}$ and $X \in \mathfrak{z}$.

(ii) $\underline{V = 0}$. *Then*

$$\begin{aligned} B_U(t) &= \sin(t)U + (1 - \cos(t))J_Y U , \\ B_X(t) &= tX \end{aligned}$$

for all $U \in \mathfrak{v}$ and $X \in \mathfrak{z}$.

(iii) $\underline{V \neq 0 \neq Y}$. *Then $\dot\gamma(t) = V(t) + Y$ with*

$$V(t) := \cos(|Y|t)V + \frac{1}{|Y|}\sin(|Y|t)J_Y V ,$$

and the span of

$$\begin{gathered} V + Y, -|Y|^2V + |V|^2Y, J_Y V, \\ \mathfrak{p} := \ker \operatorname{ad}(V) \cap \ker \operatorname{ad}(J_Y V), \\ Y^\perp, J_{Y^\perp}V, J_{Y^\perp}J_Y V \end{gathered}$$

is $\mathfrak{n}$. Any two of these vectors or vector spaces are perpendicular to each other except $J_{Y^\perp}V$ and $J_{Y^\perp}J_Y V$ (unless $K_{V,Y} = 0$). For $U \in \mathfrak{p}$ and $X \in Y^\perp$ we then have

$$B_{V+Y}(t) = tV(t) + tY ,$$

$$
\begin{aligned}
B_{-|Y|^2V+|V|^2Y}(t) &= \left(|V|^2t - \frac{1}{|Y|}\sin(|Y|t)\right)V(t) \\
&\quad + \frac{1}{|Y|^2}(1-\cos(|Y|t))\,J_YV(t) \\
&\quad + \frac{|V|^2}{|Y|^2}\left(\frac{1}{|Y|}\sin(|Y|t) - |V|^2t\right)Y\ , \\
B_{J_YV}(t) &= (1-\cos(|Y|t))V(t) + \frac{1}{|Y|}\sin(|Y|t)J_YV(t) \\
&\quad + \frac{|V|^2}{|Y|^2}(\cos(|Y|t)-1)Y\ , \\
B_U(t) &= \frac{1}{|Y|}\sin(|Y|t)U + \frac{1}{|Y|^2}(1-\cos(|Y|t))J_YU\ , \\
B_X(t) &= \left(\frac{|V|^2}{2|Y|^3}\sin(|Y|t)(\cos(|Y|t)-2) + \frac{1+|Y|^2}{2|Y|^2}t\right)X \\
&\quad + \frac{|V|^2}{2|Y|^3}(\cos(|Y|t)-1)^2KX \\
&\quad + \frac{1}{|Y|^2}\cos(|Y|t)(1-\cos(|Y|t))J_XV(t) \\
&\quad + \frac{1}{|Y|^3}\sin(|Y|t)(\cos(|Y|t)-1)J_XJ_YV(t)\ , \\
B_{J_XV}(t) &= \frac{|V|^2}{|Y|^2}\cos(|Y|t)(\cos(|Y|t)-1)X \\
&\quad + \frac{|V|^2}{|Y|^2}\sin(|Y|t)(1-\cos(|Y|t))KX \\
&\quad + \frac{1}{|Y|}\sin(|Y|t)(2\cos(|Y|t)-1)J_XV(t) \\
&\quad + \frac{1}{|Y|^2}(\cos(|Y|t)-1)(2\cos(|Y|t)+1)J_XJ_YV(t)\ , \\
B_{J_XJ_YV}(t) &= \frac{|V|^2}{|Y|}\sin(|Y|t)(\cos(|Y|t)-1)X \\
&\quad + \frac{|V|^2}{|Y|}\cos(|Y|t)(\cos(|Y|t)-1)KX \\
&\quad + (1-\cos(|Y|t))(2\cos(|Y|t)+1)J_XV(t) \\
&\quad + \frac{1}{|Y|}\sin(|Y|t)(2\cos(|Y|t)-1)J_XJ_YV(t)\ .
\end{aligned}
$$

Note that

$$
\begin{aligned}
&KX = 0 \ \textit{if and only if}\ X \in L_0\ , \\
&J_XJ_YV(t) = |Y|J_{KX}V(t)\ \textit{if}\ X \in L_k\ \textit{and}\ \mu_k = -1\ .
\end{aligned}
$$

Proof. (i) $\underline{Y=0}$. By means of Theorem 3.1.9 we have $\dot\gamma = V$. We start with

$U \in \ker \operatorname{ad}(V)$. Then

$$\nabla_V U = -\frac{1}{2}[U, V] = 0 ,$$

whence U is parallel along γ. By Theorem 3.3 we have $R_\gamma U = R_V U = 0$. This shows that the Jacobi field B_U is given by $B_U(t) = tU$.

According to 3.5 we know that $R'_\gamma = [R_\gamma, T_\gamma]$ and $T'_\gamma = 0$ with

$$T_\gamma(U + X) = \frac{1}{2}J_X V + \frac{1}{2}[U, V] .$$

Let $U \in \ker \operatorname{ad}(V)^\perp = J_3 V$ and $X := [U, V]$. Then $J_X V = -U$ and

$$\bar{\nabla}_\theta U = \nabla_\theta U + T_\gamma U = -\frac{1}{2}[U, V] + \frac{1}{2}[U, V] = 0 ,$$
$$\bar{\nabla}_\theta X = \nabla_\theta X + T_\gamma X = -\frac{1}{2}J_X V + \frac{1}{2}J_X V = 0 .$$

Thus U and X are $\bar{\nabla}_\theta$-parallel. Further, we have

$$T_\gamma U = \frac{1}{2}X , \quad T_\gamma X = -\frac{1}{2}U , \quad T_\gamma^2 U = -\frac{1}{4}U , \quad T_\gamma^2 X = -\frac{1}{4}X ,$$

and hence, using Theorem 3.3(i),

$$\bar{R}_\gamma U = R_\gamma U + T_\gamma^2 U = -\frac{3}{4}U - \frac{1}{4}U = -U ,$$
$$\bar{R}_\gamma X = R_\gamma X + T_\gamma^2 X = \frac{1}{4}X - \frac{1}{4}X = 0 .$$

Thus, in this special situation, the Jacobi equation is

$$\begin{pmatrix} z_1 \\ z_2 \end{pmatrix}'' - \begin{pmatrix} 0 & -1 \\ 1 & 0 \end{pmatrix} \begin{pmatrix} z_1 \\ z_2 \end{pmatrix}' + \begin{pmatrix} -1 & 0 \\ 0 & 0 \end{pmatrix} \begin{pmatrix} z_1 \\ z_2 \end{pmatrix} = 0 ,$$

with initial values

$$\begin{pmatrix} z_1 \\ z_2 \end{pmatrix}(0) = \begin{pmatrix} 0 \\ 0 \end{pmatrix} , \quad \begin{pmatrix} z_1 \\ z_2 \end{pmatrix}'(0) = \begin{pmatrix} 1 \\ 0 \end{pmatrix} \quad \text{for } B_U ,$$
$$\begin{pmatrix} z_1 \\ z_2 \end{pmatrix}(0) = \begin{pmatrix} 0 \\ 0 \end{pmatrix} , \quad \begin{pmatrix} z_1 \\ z_2 \end{pmatrix}'(0) = \begin{pmatrix} 0 \\ 1 \end{pmatrix} \quad \text{for } B_X ,$$

where B_U and B_X are equal to $z_1 U + z_2 X$, respectively. The above equations are

$$\begin{aligned} z_1'' + z_2' - z_1 &= 0 , \\ z_2'' - z_1' &= 0 . \end{aligned}$$

Integrating the second equation gives

$$z_2' = z_1 + z_2'(0) .$$

Inserting this into the first one yields

$$z_1'' + z_2'(0) = 0 ,$$

and therefore

$$z_1(t) = -\frac{1}{2}z_2'(0)t^2 + z_1'(0)t .$$

For z_2 we then get

$$z_2(t) = -\frac{1}{6}z_2'(0)t^3 + \frac{1}{2}z_1'(0)t^2 + z_2'(0)t .$$

Finally, taking account of the initial values then gives the expressions for B_U and B_X as stated in (i).

(ii) $V = 0$. Then $\dot\gamma = Y$. For all $X \in \mathfrak{z}$ we have $\nabla_Y X = 0$, whence X is parallel along γ. This implies $B_X(t) = tX$. According to 3.5 we have $R_\gamma' = [R_\gamma, T_\gamma]$ and $T_\gamma' = 0$ with

$$T_\gamma(U + X) = -\frac{1}{2}J_Y U .$$

Then

$$\begin{aligned}
&\bar\nabla_\theta U = \nabla_\theta U + T_\gamma U = -\frac{1}{2}J_Y U - \frac{1}{2}J_Y U = -J_Y U ,\\
&\bar\nabla_\theta J_Y U = -J_Y^2 U = U ,\\
&\bar\nabla_\theta\bar\nabla_\theta U = -U ,\\
&\bar\nabla_\theta\bar\nabla_\theta J_Y U = -J_Y U ,\\
&\bar R_\gamma U = R_\gamma U + T_\gamma^2 U = \frac{1}{4}U + \frac{1}{4}J_Y^2 U = 0 ,\\
&\bar R_\gamma J_Y U = 0 .
\end{aligned}$$

Thus, B_U is of the form

$$B_U = z_1 U + z_2 J_Y U .$$

Then

$$\begin{aligned}
\bar\nabla_\theta B_U &= z_1' U - z_1 J_Y U + z_2' J_Y U + z_2 U ,\\
\bar\nabla_\theta\bar\nabla_\theta B_U &= z_1'' U - 2z_1' J_Y U - z_1 U + z_2'' J_Y U + 2z_2' U - z_2 J_Y U .
\end{aligned}$$

So,

$$0 = \bar\nabla_\theta\bar\nabla_\theta B_U - 2T_\gamma\bar\nabla_\theta B_U + \bar R_\gamma B_U = (z_1'' + z_2')U + (z_2'' - z_1')J_Y U .$$

This implies the system of differential equations

$$z_1'' + z_2' = 0 \ , \quad z_2'' - z_1' = 0$$

with initial conditions

$$z_1(0) = 0 \ , \ z_1'(0) = 1 \ , \ z_2(0) = 0 \ , \ z_2'(0) = 0 .$$

The solutions of it are

$$z_1(t) = \sin(t) \quad , \quad z_2(t) = 1 - \cos(t) ,$$

by which (ii) is proved.

(iii) $V \neq 0 \neq Y$. The first statements in (iii) have been proved in 3.1.9 and 3.3. The formula for B_{V+Y} is a consequence of $\dot\gamma(t) = V(t) + Y$. From 3.5 we know that $R'_\gamma = [R_\gamma, T_\gamma]$ and $T'_\gamma = 0$ on $\mathfrak{n}_3 := \text{span}\{V, J_Y V, Y\}$ with

$$T_\gamma(t)(U + X) = -\frac{1}{2}J_Y U + \frac{1}{2}J_X V(t) + \frac{1}{2}[U, V(t)] \, .$$

For the sake of brevity we put

$$E_1(t) := -\frac{|Y|}{|V|}V(t) + \frac{|V|}{|Y|}Y \quad , \quad E_2(t) := \frac{1}{|V||Y|}J_Y V(t) \, .$$

E_1 and E_2 are orthonormal and, using Theorem 3.3(iii)(1), we have

$$\begin{aligned} T_\gamma E_1 &= \frac{1}{2}E_2 \quad , \quad T_\gamma E_2 = -\frac{1}{2}E_1 \, , \\ \bar{R}_\gamma E_1 &= 0 \quad , \quad \bar{R}_\gamma E_2 = -|V|^2 E_2 \, , \\ \bar{\nabla}_\theta E_1 &= 0 \quad , \quad \bar{\nabla}_\theta E_2 = 0 \, . \end{aligned}$$

Thus $B_{E_i(0)}$ is of the form $z_1 E_1 + z_2 E_2$ and the Jacobi equation is

$$\begin{pmatrix} z_1 \\ z_2 \end{pmatrix}'' - \begin{pmatrix} 0 & -1 \\ 1 & 0 \end{pmatrix}\begin{pmatrix} z_1 \\ z_2 \end{pmatrix}' + \begin{pmatrix} 0 & 0 \\ 0 & -|V|^2 \end{pmatrix}\begin{pmatrix} z_1 \\ z_2 \end{pmatrix} = 0 \, ,$$

with initial values

$$\begin{aligned} \begin{pmatrix} z_1 \\ z_2 \end{pmatrix}(0) &= \begin{pmatrix} 0 \\ 0 \end{pmatrix} , \quad \begin{pmatrix} z_1 \\ z_2 \end{pmatrix}'(0) = \begin{pmatrix} 1 \\ 0 \end{pmatrix} \quad \text{for } B_{E_1(0)} \, , \\ \begin{pmatrix} z_1 \\ z_2 \end{pmatrix}(0) &= \begin{pmatrix} 0 \\ 0 \end{pmatrix} , \quad \begin{pmatrix} z_1 \\ z_2 \end{pmatrix}'(0) = \begin{pmatrix} 0 \\ 1 \end{pmatrix} \quad \text{for } B_{E_2(0)} \, . \end{aligned}$$

Explicitly, we obtain

$$\begin{aligned} z_1'' + z_2' &= 0 \, , \\ z_2'' - z_1' - |V|^2 z_2 &= 0 \, . \end{aligned}$$

Integrating the first equation gives

$$z_1' = -z_2 + z_1'(0) \, ,$$

and inserting this into the second one implies

$$z_2'' + |Y|^2 z_2 - z_1'(0) = 0 \, .$$

Here we get the solutions

$$\begin{aligned} z_1(t) &= \frac{1}{|Y|^2}\left(\frac{1}{|Y|}\sin(|Y|t) - |V|^2 t\right) \, , \\ z_2(t) &= \frac{1}{|Y|^2}(1 - \cos(|Y|t)) \, , \\ z_1(t) &= \frac{1}{|Y|^2}(\cos(|Y|t) - 1) \, , \\ z_2(t) &= \frac{1}{|Y|}\sin(|Y|t) \, , \end{aligned}$$

for $B_{E_i(0)}$, $i = 1, 2$, respectively. Multiplication of $B_{E_1(0)}$ and $B_{E_2(0)}$ with $|V||Y|$ then gives the expressions as stated in the theorem.

Further, according to 3.5, we have $R'_\gamma = [R_\gamma, T_\gamma]$ and $T'_\gamma = 0$ on $\mathfrak{p}$ with

$$T_\gamma U = -\frac{1}{2} J_Y U \ .$$

Then, using Theorem 3.3(iii)(2), we obtain for $U \in \mathfrak{p}$,

$$\begin{array}{ll} \bar{\nabla}_\theta U = -J_Y U & , \ \bar{\nabla}_\theta J_Y U = |Y|^2 U \ , \\ \bar{\nabla}_\theta \bar{\nabla}_\theta U = -|Y|^2 U & , \ \bar{\nabla}_\theta \bar{\nabla}_\theta J_Y U = -|Y|^2 J_Y U \ , \\ \bar{R}_\gamma U = 0 & , \ \bar{R}_\gamma J_Y U = 0 \ . \end{array}$$

Thus, B_U is of the form

$$B_U = z_1 U + z_2 J_Y U \ .$$

Then

$$\begin{aligned} \bar{\nabla}_\theta B_U &= z_1' U - z_1 J_Y U + z_2' J_Y U + z_2 |Y|^2 U \ , \\ \bar{\nabla}_\theta \bar{\nabla}_\theta B_U &= z_1'' U - 2 z_1' J_Y U - z_1 |Y|^2 U + z_2'' J_Y U + 2 z_2' |Y|^2 U - z_2 |Y|^2 J_Y U \ . \end{aligned}$$

So,

$$0 = \bar{\nabla}_\theta \bar{\nabla}_\theta B_U - 2 T_\gamma \bar{\nabla}_\theta B_U + \bar{R}_\gamma B_U = (z_1'' + |Y|^2 z_2') U + (z_2'' - z_1') J_Y U \ .$$

This implies the system of differential equations

$$z_1'' + |Y|^2 z_2' = 0 \ , \ z_2'' - z_1' = 0$$

with initial conditions

$$z_1(0) = 0 \ , \ z_1'(0) = 1 \ , \ z_2(0) = 0 \ , \ z_2'(0) = 0 \ .$$

The solutions of it are

$$z_1(t) = \frac{1}{|Y|} \sin(|Y| t) \ , \ z_2(t) = \frac{1}{|Y|^2}(1 - \cos(|Y| t)) \ ,$$

by which the formula for B_U is proved.

It remains to compute the Jacobi fields for $U + X \in \mathfrak{q}$. From 3.5 we know that $R'_\gamma = [R_\gamma, T_\gamma]$ and $T'_\gamma = 0$ on $\mathfrak{q}$ for

$$T_\gamma(t)(U + X) = \frac{3}{2} J_Y U + \frac{1}{2} J_X V(t) + \frac{1}{2}[U, V(t)] \ .$$

We will use the notations as in 3.3.

We first consider the case $\mu_j \notin \{0, -1\}$. Let X be a unit vector in L_j. Then

$$T_\gamma(t) X = \frac{1}{2} J_X V(t) \ ,$$

$$
\begin{aligned}
T_\gamma(t)KX &= \frac{1}{2}J_{KX}V(t)\,, \\
T_\gamma(t)J_XV(t) &= -\frac{3}{2}J_XJ_YV(t) - \frac{1}{2}|V|^2X\,, \\
T_\gamma(t)J_{KX}V(t) &= -\frac{3}{2}J_{KX}J_YV(t) - \frac{1}{2}|V|^2KX\,, \\
T_\gamma(t)J_XJ_YV(t) &= \frac{3}{2}|Y|^2J_XV(t) - \frac{1}{2}|V|^2|Y|KX\,, \\
T_\gamma(t)J_{KX}J_YV(t) &= \frac{3}{2}|Y|^2J_{KX}V(t) - \frac{1}{2}\mu_j|V|^2|Y|X\,.
\end{aligned}
$$

Therefore,

$$
\mathfrak{q}_j(X) := \operatorname{span}\{X, KX, J_XV(t), J_{KX}V(t), J_XJ_YV(t), J_{KX}J_YV(t)\}
$$

is invariant under the action of $T_\gamma(t)$. Further, we easily verify that

$$
X\,,\; KX\,,\; J_XV(t)\,,\; J_{KX}V(t)\,,\; J_XJ_YV(t)\,,\; J_{KX}J_YV(t)
$$

are $\bar\nabla_\theta$-parallel. Unfortunately, the vector fields $J_XV(t)$ and $J_{KX}J_YV(t)$ (and also $J_{KX}V(t)$ and $J_XJ_YV(t)$, respectively) are not orthogonal to each other, since

$$
<J_XV(t), J_{KX}J_YV(t)> = |V|^2|Y|<K^2X, X> = |V|^2|Y|\mu_j \neq 0\,.
$$

But we get a $\bar\nabla_\theta$-parallel orthonormal frame field $E_1,\ldots,E_6$ of $\mathfrak{q}_j(X)$ along γ by defining

$$
\begin{aligned}
E_1(t) &:= X\,, \\
E_2(t) &:= \frac{1}{\sqrt{-\mu_j}}KX\,, \\
E_3(t) &:= \frac{1}{|V|}J_XV(t)\,, \\
E_4(t) &:= \frac{1}{|V|\sqrt{-\mu_j}}J_{KX}V(t)\,, \\
E_5(t) &:= \frac{1}{|V||Y|\sqrt{1+\mu_j}}(J_XJ_YV(t) - |Y|J_{KX}V(t))\,, \\
E_6(t) &:= \frac{1}{|V||Y|\sqrt{-\mu_j}\sqrt{1+\mu_j}}(J_{KX}J_YV(t) - \mu_j|Y|J_XV(t))\,.
\end{aligned}
$$

We continue the computations to get

$$
\begin{aligned}
&T_\gamma(t)(J_XJ_YV(t) - |Y|J_{KX}V(t)) \\
&\quad= \frac{3}{2}|Y|(J_{KX}J_YV(t) - \mu_j|Y|J_XV(t)) + \frac{3}{2}|Y|^2(1+\mu_j)J_XV(t)\,, \\
&T_\gamma(t)(J_{KX}J_YV(t) - \mu_j|Y|J_XV(t)) \\
&\quad= \frac{3}{2}\mu_j|Y|(J_XJ_YV(t) - |Y|J_{KX}V(t)) + \frac{3}{2}|Y|^2(1+\mu_j)J_{KX}V(t)\,,
\end{aligned}
$$

and

$$\begin{aligned}
\bar{R}_\gamma(t)X &= 0\,, \\
\bar{R}_\gamma(t)KX &= 0\,, \\
\bar{R}_\gamma(t)J_XV(t) &= -(1+|Y|^2)J_XV(t)\,, \\
\bar{R}_\gamma(t)J_{KX}V(t) &= -(1+|Y|^2)J_{KX}V(t)\,, \\
\bar{R}_\gamma(t)(J_XJ_YV(t)-|Y|J_{KX}V(t)) &= -2|Y|^2(J_XJ_YV(t)-|Y|J_{KX}V(t))\,, \\
\bar{R}_\gamma(t)(J_{KX}J_YV(t)-\mu_j|Y|J_XV(t)) &= -2|Y|^2(J_{KX}J_YV(t)-\mu_j|Y|J_XV(t))\,.
\end{aligned}$$

So, with respect to $E_1,\ldots,E_6$, the Jacobi equation along γ, restricted to $\mathfrak{q}_j(X)$, becomes

$$z''-Pz'+Qz=0$$

with $z:=(z_1,z_2,z_3,z_4,z_5,z_6)^T$,

$$P:=\begin{pmatrix}
0 & 0 & -|V| & 0 & 0 & 0 \\
0 & 0 & 0 & -|V| & 0 & 0 \\
|V| & 0 & 0 & 3|Y|\sqrt{-\mu_j} & 3|Y|\sqrt{1+\mu_j} & 0 \\
0 & |V| & -3|Y|\sqrt{-\mu_j} & 0 & 0 & 3|Y|\sqrt{1+\mu_j} \\
0 & 0 & -3|Y|\sqrt{1+\mu_j} & 0 & 0 & -3|Y|\sqrt{-\mu_j} \\
0 & 0 & 0 & -3|Y|\sqrt{1+\mu_j} & 3|Y|\sqrt{-\mu_j} & 0
\end{pmatrix}$$

and

$$Q:=\begin{pmatrix}
0 & 0 & 0 & 0 & 0 & 0 \\
0 & 0 & 0 & 0 & 0 & 0 \\
0 & 0 & -(1+|Y|^2) & 0 & 0 & 0 \\
0 & 0 & 0 & -(1+|Y|^2) & 0 & 0 \\
0 & 0 & 0 & 0 & -2|Y|^2 & 0 \\
0 & 0 & 0 & 0 & 0 & -2|Y|^2
\end{pmatrix}.$$

This gives the system of differential equations

$$\begin{aligned}
0 &= z_1''+|V|z_3'\,, \\
0 &= z_2''+|V|z_4'\,, \\
0 &= z_3''-|V|z_1'-3|Y|\sqrt{-\mu_j}z_4'-3|Y|\sqrt{1+\mu_j}z_5'-(1+|Y|^2)z_3\,, \\
0 &= z_4''-|V|z_2'+3|Y|\sqrt{-\mu_j}z_3'-3|Y|\sqrt{1+\mu_j}z_6'-(1+|Y|^2)z_4\,, \\
0 &= z_5''+3|Y|\sqrt{1+\mu_j}z_3'+3|Y|\sqrt{-\mu_j}z_6'-2|Y|^2z_5\,, \\
0 &= z_6''+3|Y|\sqrt{1+\mu_j}z_4'-3|Y|\sqrt{-\mu_j}z_5'-2|Y|^2z_6\,.
\end{aligned}$$

Integration of the first two equations yields

$$\begin{aligned}
z_1' &= -|V|z_3+z_1'(0)\,, \\
z_2' &= -|V|z_4+z_2'(0)\,.
\end{aligned}$$

Inserting this into the equations for z_3 and z_4 provides

$$\begin{aligned}
0 &= z_3''-2|Y|^2z_3-3|Y|\sqrt{-\mu_j}z_4'-3|Y|\sqrt{1+\mu_j}z_5'-|V|z_1'(0)\,, \\
0 &= z_4''-2|Y|^2z_4+3|Y|\sqrt{-\mu_j}z_3'-3|Y|\sqrt{1+\mu_j}z_6'-|V|z_2'(0)\,.
\end{aligned}$$

We differentiate the first one and insert the last equation for z_4 and the one for z_5 to obtain

$$z_3''' + 7|Y|^2 z_3' - 6|Y|^3\sqrt{-\mu_j}\, z_4 - 6|Y|^3\sqrt{1+\mu_j}\, z_5 - 3|V||Y|\sqrt{-\mu_j}\, z_2'(0) = 0 .$$

Differentiating again and inserting the last second order equation for z_3 gives

$$z_3'''' + 5|Y|^2 z_3'' + 4|Y|^4 z_3 + 2|V||Y|^2 z_1'(0) = 0 .$$

As

$$z^4 + 5|Y|^2 z^2 + 4|Y|^4 = (z^2 + |Y|^2)(z^2 + 4|Y|^2) ,$$

we have the general solution

$$z_3(t) = \alpha\cos(|Y|t) + \beta\sin(|Y|t) + \delta\cos(2|Y|t) + \eta\sin(2|Y|t) - \frac{|V|z_1'(0)}{2|Y|^2} .$$

We now compute analogously for z_4 to obtain

$$z_4''' + 7|Y|^2 z_4' + 6|Y|^3\sqrt{-\mu_j}\, z_3 - 6|Y|^3\sqrt{1+\mu_j}\, z_6 + 3|V||Y|\sqrt{-\mu_j}\, z_1'(0) = 0 .$$

Differentiating again and inserting the last second order equation for z_4 leads to

$$z_4'''' + 5|Y|^2 z_4'' + 4|Y|^4 z_4 + 2|V||Y|^2 z_2'(0) = 0 .$$

So, we have the general solution

$$z_4(t) = \tilde\alpha\cos(|Y|t) + \tilde\beta\sin(|Y|t) + \tilde\delta\cos(2|Y|t) + \tilde\eta\sin(2|Y|t) - \frac{|V|z_2'(0)}{2|Y|^2} .$$

We now have to take account of the initial values.

For $B_{E_1(0)}$: The initial values for z_3 are

$$\begin{aligned} 0 = z_3(0) &= \alpha + \delta - \frac{|V|}{2|Y|^2} , \\ 0 = z_3'(0) &= |Y|(\beta + 2\eta) , \\ |V| = z_3''(0) &= -|Y|^2(\alpha + 4\delta) , \\ 0 = z_3'''(0) &= -|Y|^3(\beta + 8\eta) . \end{aligned}$$

A simple computation gives

$$\alpha = \frac{|V|}{|Y|^2} , \ \beta = 0 , \ \delta = -\frac{|V|}{2|Y|^2} , \ \eta = 0 .$$

Inserting this in the general form of the solution yields

$$z_3(t) = \frac{|V|}{|Y|^2}\left(\cos(|Y|t) - \frac{1}{2}\cos(2|Y|t) - \frac{1}{2}\right) .$$

From $z_1' = -|V|z_3 + 1$ and $z_1(0) = 0$ we then deduce

$$z_1(t) = \frac{|V|^2}{|Y|^3}\left(\frac{1}{4}\sin(2|Y|t) - \sin(|Y|t)\right) + \frac{1+|Y|^2}{2|Y|^2}t .$$

For z_4 we have the initial values

$$\begin{aligned} 0 = z_4(0) &= \tilde\alpha + \tilde\delta , \\ 0 = z_4'(0) &= |Y|(\tilde\beta + 2\tilde\eta) , \\ 0 = z_4''(0) &= -|Y|^2(\tilde\alpha + 4\tilde\delta) , \\ -3|V||Y|\sqrt{-\mu_j} = z_4'''(0) &= -|Y|^3(\tilde\beta + 8\tilde\eta) . \end{aligned}$$

A straightforward computation shows that

$$\tilde\alpha = 0 , \ \tilde\beta = -\frac{|V|}{|Y|^2}\sqrt{-\mu_j} , \ \tilde\delta = 0 , \ \tilde\eta = \frac{|V|}{2|Y|^2}\sqrt{-\mu_j} .$$

Inserting this in the general solution yields now

$$z_4(t) = \frac{|V|}{|Y|^2}\sqrt{-\mu_j}\left(\frac{1}{2}\sin(2|Y|t) - \sin(|Y|t)\right) .$$

From $z_2' = -|V|z_4$ and $z_2(0) = 0$ we then get

$$z_2(t) = \frac{|V|^2}{|Y|^3}\sqrt{-\mu_j}\left(\frac{1}{4}\cos(2|Y|t) - \cos(|Y|t) + \frac{3}{4}\right) .$$

Next, for z_5 and z_6 we find

$$\begin{aligned} z_5(t) &= \frac{1}{6|Y|^3\sqrt{1+\mu_j}}\left(z_3'''(t) + 7|Y|^2 z_3'(t) - 6|Y|^3\sqrt{-\mu_j}\, z_4(t)\right) \\ &= \frac{|V|}{|Y|^2}\sqrt{1+\mu_j}\left(\frac{1}{2}\sin(2|Y|t) - \sin(|Y|t)\right) . \\ z_6(t) &= \frac{1}{6|Y|^3\sqrt{1+\mu_j}}\left(z_4'''(t) + 7|Y|^2 z_4'(t) + 6|Y|^3\sqrt{-\mu_j}\, z_3(t) + 3|V||Y|\sqrt{-\mu_j}\right) \\ &= 0 . \end{aligned}$$

Summing up, we get

$$\begin{aligned} B_X(t) &= \left(\frac{|V|^2}{|Y|^3}\left(\frac{1}{4}\sin(2|Y|t) - \sin(|Y|t)\right) + \frac{1+|Y|^2}{2|Y|^2}t\right) E_1(t) \\ &\quad + \frac{|V|^2}{|Y|^3}\sqrt{-\mu_j}\left(\frac{1}{4}\cos(2|Y|t) - \cos(|Y|t) + \frac{3}{4}\right) E_2(t) \\ &\quad + \frac{|V|}{|Y|^2}\left(\cos(|Y|t) - \frac{1}{2}\cos(2|Y|t) - \frac{1}{2}\right) E_3(t) \\ &\quad + \frac{|V|}{|Y|^2}\sqrt{-\mu_j}\left(\frac{1}{2}\sin(2|Y|t) - \sin(|Y|t)\right) E_4(t) \\ &\quad + \frac{|V|}{|Y|^2}\sqrt{1+\mu_j}\left(\frac{1}{2}\sin(2|Y|t) - \sin(|Y|t)\right) E_5(t) \\ &= \left(\frac{|V|^2}{|Y|^3}\left(\frac{1}{4}\sin(2|Y|t) - \sin(|Y|t)\right) + \frac{1+|Y|^2}{2|Y|^2}t\right) X \\ &\quad + \frac{|V|^2}{|Y|^3}\left(\frac{1}{4}\cos(2|Y|t) - \cos(|Y|t) + \frac{3}{4}\right) KX \end{aligned}$$

$$
\begin{aligned}
&+\frac{1}{|Y|^2}\left(\cos(|Y|t)-\frac{1}{2}\cos(2|Y|t)-\frac{1}{2}\right)J_X V(t)\\
&+\frac{1}{|Y|^3}\left(\frac{1}{2}\sin(2|Y|t)-\sin(|Y|t)\right)J_X J_Y V(t)\\
=\ &\left(\frac{|V|^2}{2|Y|^3}\sin(|Y|t)(\cos(|Y|t)-2)+\frac{1+|Y|^2}{2|Y|^2}t\right)X\\
&+\frac{|V|^2}{2|Y|^3}(\cos(|Y|t)-1)^2 KX\\
&+\frac{1}{|Y|^2}\cos(|Y|t)(1-\cos(|Y|t))J_X V(t)\\
&+\frac{1}{|Y|^3}\sin(|Y|t)(\cos(|Y|t)-1)J_X J_Y V(t)\ .
\end{aligned}
$$

For $B_{E_3(0)}$: For z_3 we have the initial values

$$z_3(0)=0\ ,\ z_3'(0)=1\ ,\ z_3''(0)=0\ ,\ z_3'''(0)=-7|Y|^2\ .$$

Here we obtain

$$\alpha=0\ ,\ \beta=-\frac{1}{|Y|}\ ,\ \delta=0\ ,\ \eta=\frac{1}{|Y|}\ .$$

Inserting this in the general solution gives

$$z_3(t)=\frac{1}{|Y|}(\sin(2|Y|t)-\sin(|Y|t))\ .$$

From $z_1'=-|V|z_3$ and $z_1(0)=0$ we deduce

$$z_1(t)=\frac{|V|}{|Y|^2}\left(\frac{1}{2}\cos(2|Y|t)-\cos(|Y|t)+\frac{1}{2}\right)\ .$$

The initial values for z_4 are

$$z_4(0)=0\ ,\ z_4'(0)=0\ ,\ z_4''(0)=-3|Y|\sqrt{-\mu_j}\ ,\ z_4'''(0)=0\ .$$

A straightforward computation shows that

$$\tilde{\alpha}=-\frac{1}{|Y|}\sqrt{-\mu_j}\ ,\ \tilde{\beta}=0\ ,\ \tilde{\delta}=\frac{1}{|Y|}\sqrt{-\mu_j}\ ,\ \tilde{\eta}=0\ .$$

Inserting this in the general solution gives

$$z_4(t)=\frac{1}{|Y|}\sqrt{-\mu_j}(\cos(2|Y|t)-\cos(|Y|t))\ .$$

From $z_2'=-|V|z_4$ and $z_2(0)=0$ we get

$$z_2(t)=\frac{|V|}{|Y|^2}\sqrt{-\mu_j}\left(\sin(|Y|t)-\frac{1}{2}\sin(2|Y|t)\right)\ .$$

Eventually, for z_5 and z_6 we obtain

$$\begin{aligned}
z_5(t) &= \frac{1}{6|Y|^3\sqrt{1+\mu_j}}(z_3'''(t) + 7|Y|^2 z_3'(t) - 6|Y|^3\sqrt{-\mu_j} z_4(t)) \\
&= \frac{1}{|Y|}\sqrt{1+\mu_j}(\cos(2|Y|t) - \cos(|Y|t)) \, , \\
z_6(t) &= \frac{1}{6|Y|^3\sqrt{1+\mu_j}}(z_4'''(t) + 7|Y|^2 z_4'(t) + 6|Y|^3\sqrt{-\mu_j} z_3(t)) \\
&= 0 \, .
\end{aligned}$$

Summing up, we get

$$\begin{aligned}
B_{E_3(0)}(t) &= \frac{|V|}{|Y|^2}\left(\frac{1}{2}\cos(2|Y|t) - \cos(|Y|t) + \frac{1}{2}\right) E_1(t) \\
&+ \frac{|V|}{|Y|^2}\sqrt{-\mu_j}\left(\sin(|Y|t) - \frac{1}{2}\sin(2|Y|t)\right) E_2(t) \\
&+ \frac{1}{|Y|}(\sin(2|Y|t) - \sin(|Y|t))E_3(t) \\
&+ \frac{1}{|Y|}\sqrt{-\mu_j}(\cos(2|Y|t) - \cos(|Y|t))E_4(t) \\
&+ \frac{1}{|Y|}\sqrt{1+\mu_j}(\cos(2|Y|t) - \cos(|Y|t))E_5(t) \, ,
\end{aligned}$$

and hence

$$\begin{aligned}
B_{J_X V}(t) &= \frac{|V|^2}{|Y|^2}\left(\frac{1}{2}\cos(2|Y|t) - \cos(|Y|t) + \frac{1}{2}\right) X \\
&+ \frac{|V|^2}{|Y|^2}\left(\sin(|Y|t) - \frac{1}{2}\sin(2|Y|t)\right) KX \\
&+ \frac{1}{|Y|}(\sin(2|Y|t) - \sin(|Y|t))J_X V(t) \\
&+ \frac{1}{|Y|^2}(\cos(2|Y|t) - \cos(|Y|t))J_X J_Y V(t) \\
&= \frac{|V|^2}{|Y|^2}\cos(|Y|t)(\cos(|Y|t) - 1)X \\
&+ \frac{|V|^2}{|Y|^2}\sin(|Y|t)(1 - \cos(|Y|t))KX \\
&+ \frac{1}{|Y|}\sin(|Y|t)(2\cos(|Y|t) - 1)J_X V(t) \\
&+ \frac{1}{|Y|^2}(\cos(|Y|t) - 1)(2\cos(|Y|t) + 1)J_X J_Y V(t) \, .
\end{aligned}$$

For $B_{E_5(0)}$: For z_3 we have the initial values

$$z_3(0) = 0 \, , \; z_3'(0) = 0 \, , \; z_3''(0) = 3|Y|\sqrt{1+\mu_j} \, , \; z_3'''(0) = 0 \, .$$

This implies

$$\alpha = \frac{1}{|Y|}\sqrt{1+\mu_j}\ ,\ \beta = 0\ ,\ \delta = -\frac{1}{|Y|}\sqrt{1+\mu_j}\ ,\ \eta = 0\ .$$

Inserting this in the general form of the solution provides

$$z_3(t) = \frac{1}{|Y|}\sqrt{1+\mu_j}(\cos(|Y|t) - \cos(2|Y|t))\ .$$

From $z_1' = -|V|z_3$ and $z_1(0) = 0$ we deduce

$$z_1(t) = \frac{|V|}{|Y|^2}\sqrt{1+\mu_j}\left(\frac{1}{2}\sin(2|Y|t) - \sin(|Y|t)\right)\ .$$

The initial values for z_4 are

$$z_4(0) = 0\ ,\ z_4'(0) = 0\ ,\ z_4''(0) = 0\ ,\ z_4'''(0) = 0\ .$$

Thus

$$z_4 = 0\ ,$$

and from $z_2' = -|V|z_4$ and $z_2(0) = 0$ we get

$$z_2 = 0\ .$$

Eventually, for z_5 and z_6 we get

$$\begin{aligned} z_5(t) &= \frac{1}{6|Y|^3\sqrt{1+\mu_j}}(z_3'''(t) + 7|Y|^2 z_3'(t)) \\ &= \frac{1}{|Y|}(\sin(2|Y|t) - \sin(|Y|t))\ , \\ z_6(t) &= \frac{1}{6|Y|^3\sqrt{1+\mu_j}}6|Y|^3\sqrt{-\mu_j}z_3(t) \\ &= \frac{1}{|Y|}\sqrt{-\mu_j}(\cos(|Y|t) - \cos(2|Y|t))\ . \end{aligned}$$

Summing up, we get

$$\begin{aligned} B_{E_s(0)}(t) &= \frac{|V|}{|Y|^2}\sqrt{1+\mu_j}\left(\frac{1}{2}\sin(2|Y|t) - \sin(|Y|t)\right)E_1(t) \\ &\quad + \frac{1}{|Y|}\sqrt{1+\mu_j}(\cos(|Y|t) - \cos(2|Y|t))E_3(t) \\ &\quad + \frac{1}{|Y|}(\sin(2|Y|t) - \sin(|Y|t))E_5(t) \\ &\quad + \frac{1}{|Y|}\sqrt{-\mu_j}(\cos(|Y|t) - \cos(2|Y|t))E_6(t)\ , \end{aligned}$$

or equivalently,

$$\begin{aligned}
B_{J_X J_Y V - |Y| J_{KX} V}(t) &= \frac{|V|^2}{|Y|}(1+\mu_j)\left(\frac{1}{2}\sin(2|Y|t) - \sin(|Y|t)\right) X \\
&\quad + (1+\mu_j)(\cos(|Y|t) - \cos(2|Y|t)) J_X V(t) \\
&\quad + \frac{1}{|Y|}(\sin(2|Y|t) - \sin(|Y|t))(J_X J_Y V(t) - |Y| J_{KX} V(t)) \\
&\quad + \frac{1}{|Y|}(\cos(|Y|t) - \cos(2|Y|t))(J_{KX} J_Y V(t) - \mu_j |Y| J_X V(t)) \\
&= \frac{|V|^2}{|Y|}(1+\mu_j)\left(\frac{1}{2}\sin(2|Y|t) - \sin(|Y|t)\right) X \\
&\quad + (\cos(|Y|t) - \cos(2|Y|t)) J_X V(t) \\
&\quad + (\sin(|Y|t) - \sin(2|Y|t)) J_{KX} V(t) \\
&\quad + \frac{1}{|Y|}(\sin(2|Y|t) - \sin(|Y|t)) J_X J_Y V(t) \\
&\quad + \frac{1}{|Y|}(\cos(|Y|t) - \cos(2|Y|t)) J_{KX} J_Y V(t) \, .
\end{aligned}$$

As (replace X by $|Y|KX$ in the case $B_{E_3(0)}$)

$$\begin{aligned}
B_{|Y| J_{KX} V}(t) &= \frac{|V|^2}{|Y|}\left(\frac{1}{2}\cos(2|Y|t) - \cos(|Y|t) + \frac{1}{2}\right) KX \\
&\quad + \frac{|V|^2}{|Y|}\mu_j \left(\sin(|Y|t) - \frac{1}{2}\sin(2|Y|t)\right) X \\
&\quad + (\sin(2|Y|t) - \sin(|Y|t)) J_{KX} V(t) \\
&\quad + \frac{1}{|Y|}(\cos(2|Y|t) - \cos(|Y|t)) J_{KX} J_Y V(t) \, ,
\end{aligned}$$

we finally get

$$\begin{aligned}
B_{J_X J_Y V}(t) &= \frac{|V|^2}{|Y|}\left(\frac{1}{2}\sin(2|Y|t) - \sin(|Y|t)\right) X \\
&\quad + \frac{|V|^2}{|Y|}\left(\frac{1}{2}\cos(2|Y|t) - \cos(|Y|t) + \frac{1}{2}\right) KX \\
&\quad + (\cos(|Y|t) - \cos(2|Y|t)) J_X V(t) \\
&\quad + \frac{1}{|Y|}(\sin(2|Y|t) - \sin(|Y|t)) J_X J_Y V(t) \\
&= \frac{|V|^2}{|Y|}\sin(|Y|t)(\cos(|Y|t) - 1) X \\
&\quad + \frac{|V|^2}{|Y|}\cos(|Y|t)(\cos(|Y|t) - 1) KX \\
&\quad + (1 - \cos(|Y|t))(2\cos(|Y|t) + 1) J_X V(t) \\
&\quad + \frac{1}{|Y|}\sin(|Y|t)(2\cos(|Y|t) - 1) J_X J_Y V(t) \, .
\end{aligned}$$

The cases $\mu_0 = 0$ and $\mu_k = -1$ can be treated as follows. First, if $\mu_0 = 0$, then $KX = 0$ for all $X \in L_0$. For a unit vector $X \in L_0$ define

$$\mathfrak{q}_0(X) := \text{span}\{X, J_X V(t), J_X J_Y V(t)\} .$$

The Jacobi equation along γ, restricted to $\mathfrak{q}_0(X)$, is then given by the subsystem of the system of equations of the preceding general case obtained by considering the $\bar{\nabla}_\theta$-parallel orthonormal frame field E_1, E_3, E_5 of $\mathfrak{q}_0(X)$ along γ. Secondly, if $\mu_k = -1$, then $J_X J_Y V(t) = |Y| J_{KX} V(t)$ for all $X \in L_k$. For a unit vector $X \in L_k$ define

$$\mathfrak{q}_k(X) := \text{span}\{X, KX, J_X V(t), J_{KX} V(t)\} .$$

Here, the Jacobi equation along γ, restricted to $\mathfrak{q}_k(X)$, is given by the subsystem of the system of equations of the preceding general case obtained by considering the $\bar{\nabla}_\theta$-parallel orthonormal frame field E_1, E_2, E_3, E_4 of $\mathfrak{q}_k(X)$ along γ. In both special cases the solutions are as in the general case with $KX = 0$ or $J_X J_Y V(t) = |Y| J_{KX} V(t)$, respectively.

By this the theorem is now proved. □

3.8 Conjugate points

As a first application of Theorem 3.7 we will now compute some conjugate points in generalized Heisenberg groups. We denote the exponential map at the identity e by $\exp_e$ and its differential by $\exp_{e*}$.

Theorem *Let $V + Y \in \mathfrak{n}$ be a unit vector and $\gamma : \mathbb{R} \to N$ the geodesic in N with $\gamma(0) = e$ and $\dot\gamma(0) = V + Y$. We have*

(i) $\underline{Y = 0}$. There are no conjugate points along γ.

(ii) $\underline{V = 0}$. The conjugate points along γ are at $t \in 2\pi\mathbb{Z}^*$. The multiplicity of these conjugate points is n and the kernel of $\exp_{e*}$ at tY is $\mathfrak{v}$.

(iii) $\underline{V \neq 0 \neq Y}$.

(1) Every t with $|Y|t \in 2\pi\mathbb{Z}^*$ determines a conjugate point along γ with multiplicity $n-1$, and the kernel of $\exp_{e*}$ at $t(V+Y)$ is the orthogonal complement $V^\perp$ of $\mathbb{R}V$ in $\mathfrak{v}$.

(2) Every t satisfying the equation

$$\frac{1}{|V|^2} = \frac{|Y|t}{2} \cot\left(\frac{|Y|t}{2}\right)$$

determines a conjugate point along γ. The first positive t with this property is greater than $2\pi/|Y|$.

Proof. (i) $\underline{Y=0}$. Let $U+X \in \mathfrak{n}$ be a non-zero vector. According to Theorem 3.7 we have

$$B_{U+X}(t) = tU + \frac{1}{2}t^2 J_X V + \left(t - \frac{1}{6}t^3\right) X + \frac{1}{2}t^2[U,V] .$$

If U and $J_X V$ are linearly independent, then $B_{U+X}(t)$ vanishes if and only if $t=0$. Now suppose that U and $J_X V$ are linearly dependent. If $J_X V = 0$, then $B_{U+X}(t) = 0$ holds only if $t=0$. If $J_X V \neq 0$, then $U = \alpha J_X V$ for some $\alpha \in \mathbb{R}$, and we obtain

$$B_{U+X}(t) = \left(\alpha + \frac{1}{2}t\right) t J_X V + \left(1 - \frac{1}{6}t^2 - \frac{1}{2}\alpha t\right) tX .$$

If $B_{U+X}(t) = 0$, then the $J_X V$-component gives $t=0$ or $t=-2\alpha$. If $t=-2\alpha$, the X-component is $-2(1+\alpha^2/3)\alpha$ and vanishes precisely when $\alpha = 0$. So we conclude that $B_{U+X}(t) = 0$ if and only if $t=0$, whence there are no conjugate points along γ.

(ii) $\underline{V=0}$. Let $U+X \in \mathfrak{n}$ be a non-zero vector. Then

$$B_{U+X}(t) = \sin(t)U + (1-\cos(t))J_Y U + tX .$$

Here one sees readily that $B_{U+X}(t)$ vanishes at some $t \neq 0$ if and only if $X=0$ and $t \in 2\pi\mathbb{Z}^*$.

(iii) $\underline{V \neq 0 \neq Y}$. Let $U+X \in \mathfrak{n}$ be non-zero. We may assume that $U+X$ is orthogonal to $V+Y$, because otherwise $B_{U+X}(t)$ vanishes only if $t=0$. If $U+X$ is in $\mathfrak{n}_3$, $\mathfrak{p}$ or $\mathfrak{q}$, then, according to Theorem 3.7, $B_{U+X}(t)$ is in $\mathfrak{n}_3$, $\mathfrak{p}$ and $\mathfrak{q}$, respectively, for all t. Thus, in order to find the conjugate points along γ it suffices to investigate the Jacobi fields B_{U+X} with initial values $B_{U+X}(0) = 0$ and $B'_{U+X}(0)$ in $\mathfrak{n}_3$, $\mathfrak{p}$ or $\mathfrak{q}$. The simplest case to deal with is when $U+X \in \mathfrak{p}$ (so $X=0$). By means of Theorem 3.7 we have

$$B_U(t) = \frac{1}{|Y|}\sin(|Y|t)U + \frac{1}{|Y|^2}(1-\cos(|Y|t))J_Y U .$$

Clearly, $B_U(t) = 0$ precisely if $|Y|t \in 2\pi\mathbb{Z}$.

Next, if $U+X \in \mathfrak{n}_3$, we put

$$E_1(t) := -\frac{|Y|}{|V|}V(t) + \frac{|V|}{|Y|}Y \ , \ E_2(t) := \frac{1}{|V||Y|}J_Y V(t)$$

with

$$V(t) := \cos(|Y|t)V + \frac{1}{|Y|}\sin(|Y|t)J_Y V ,$$

and

$$U + X = \alpha E_1(0) + \beta E_2(0)$$

with some $\alpha, \beta \in \mathbb{R}$, not both equal to zero (recall that $U+X$ is a non-zero vector orthogonal to $V+Y$). From Theorem 3.7 we then obtain

$$\begin{aligned} B_{U+X}(t) &= \frac{1}{|Y|^3}(\alpha(\sin(|Y|t) - |V|^2|Y|t) + \beta|Y|(\cos(|Y|t)-1))E_1(t) \\ &\quad + \frac{1}{|Y|^2}(\alpha(1-\cos(|Y|t)) + \beta|Y|\sin(|Y|t))E_2(t) . \end{aligned}$$

The determinant of the matrix

$$\begin{pmatrix} \sin(|Y|t) - |V|^2|Y|t & |Y|(\cos(|Y|t) - 1) \\ |Y|(1 - \cos(|Y|t)) & |Y|^2 \sin(|Y|t) \end{pmatrix}$$

is equal to

$$\begin{aligned} & |Y|^2(2(1 - \cos(|Y|t)) - |V|^2|Y|t\sin(|Y|t)) \\ = \; & |Y|^2 \left(4\sin^2\left(\frac{|Y|t}{2}\right) - 2|V|^2|Y|t\sin\left(\frac{|Y|t}{2}\right)\cos\left(\frac{|Y|t}{2}\right)\right) , \end{aligned}$$

and vanishes precisely if

$$|Y|t \in 2\pi\mathbb{Z} \quad \text{or} \quad \frac{1}{|V|^2} = \frac{|Y|t}{2}\cot\left(\frac{|Y|t}{2}\right) ,$$

Note that the first condition implies $\alpha = 0$ and hence $U + X = \beta J_Y V$. Further, since $|V|^2 < 1$, the first positive t satisfying the second equation is subjected to the inequality $|Y|t > 2\pi$. So we have obtained all conjugate points along γ arising from initial values in $\mathfrak{n}_3 \cap (V+Y)^\perp$.

Finally, if $|Y|t \in 2\pi\mathbb{Z}$ and $X \in Y^\perp$, then

$$B_X(t) = \frac{1+|Y|^2}{2|Y|^2}tX , \quad B_{J_X V}(t) = 0 , \quad B_{J_X J_Y V}(t) = 0 .$$

This shows that the multiplicity of the conjugate points with $|Y|t \in 2\pi\mathbb{Z}^*$ is $n-1$, and the kernel of $\exp_{e*}$ at $t(V+Y)$ is

$$\mathbb{R}J_Y V \oplus \mathfrak{p} \oplus (\mathfrak{q} \cap \mathfrak{v}) = V^\perp .$$

By this also statement (iii) is proved. □

Theoretically the same method may be used to decide whether there arise further conjugate points from initial values in $\mathfrak{q}$, but the explicit computations become much more complicated. We did not pursue this work in order to have a complete answer. On the other hand, $\mathfrak{q}$ is trivial if $\dim \mathfrak{z} = 1$ and then we obtain as a corollary a complete classification of the conjugate points in Heisenberg groups.

Corollary *Let N be a $(2k+1)$-dimensional Heisenberg group, $V+Y \in \mathfrak{n}$ a unit vector and $\gamma : \mathbb{R} \to N$ the geodesic in N with $\gamma(0) = e$ and $\dot\gamma(0) = V+Y$. We have*

(i) $\underline{Y=0}$. There are no conjugate points along γ.

(ii) $\underline{V=0}$. The conjugate points along γ are at $t \in 2\pi\mathbb{Z}^*$, and their multiplicity is $2k$.

(iii) $\underline{V \neq 0 \neq Y}$. The conjugate points along γ are at all t satisfying

$$|Y|t \in 2\pi\mathbb{Z}^* \quad \text{or} \quad \frac{1}{|V|^2} = \frac{|Y|t}{2}\cot\left(\frac{|Y|t}{2}\right) .$$

In the first case the multiplicity of the conjugate point is $2k-1$, in the second case it is one. The first conjugate point is at $2\pi/|Y|$.

Partial information about conjugate points in generalized Heisenberg groups has also been obtained by J. Boggino [Bog].

3.9 Principal curvatures of geodesic spheres

The main result of this section says that the principal curvatures of small geodesic spheres in generalized Heisenberg groups are the same at antipodal points. To prove this, we start with the following

Lemma *Let M be a Riemannian manifold with Levi Civita connection ∇, $p \in M$, $\gamma : [0,b] \to M$ a geodesic parametrized by arc length and with $\gamma(0) = p$, and $r \in]0,b]$ so that the geodesic sphere $G_p(r)$ centered at p and with radius r is a hypersurface of M. Denote by ξ the "outward" unit normal field of $G_p(r)$ and by A the shape operator of $G_p(r)$ with respect to ξ, that is, $Av = -\nabla_v \xi$ for all $v \in TG_p(r)$. Let T_γ be a parallel skew-symmetric tensor field along γ. As in 3.7 we define*

$$\bar{\nabla}_\partial := \nabla_\partial + T_\gamma \quad \text{and} \quad \bar{R}_\gamma := R_\gamma + T_\gamma^2 \ .$$

Let $\bar{D}$ be the solution of the $\mathrm{End}(\dot\gamma^\perp)$*-valued second order equation*

$$\bar{\nabla}_\partial \bar{\nabla}_\partial \bar{D} - 2T_\gamma \bar{\nabla}_\partial \bar{D} + \bar{R}_\gamma \bar{D} = 0$$

with initial values

$$\bar{D}(0) = 0 \quad \text{and} \quad (\bar{\nabla}_\partial \bar{D})(0) = id_{\dot\gamma(0)^\perp} \ .$$

Then

$$A_{\gamma(r)} = -(\bar{\nabla}_\partial \bar{D})(r) \circ \bar{D}^{-1}(r) + T_\gamma(r) \ .$$

Proof. Let $\bar{E}$ be a $\bar{\nabla}_\partial$-parallel vector field along γ which is perpendicular to $\dot\gamma$. For $Y := \bar{D}\bar{E}$ we then have

$$\begin{aligned} \nabla_\partial \nabla_\partial Y + R_\gamma Y &= \bar{\nabla}_\partial \bar{\nabla}_\partial (\bar{D}\bar{E}) - 2T_\gamma \bar{\nabla}_\partial (\bar{D}\bar{E}) + \bar{R}_\gamma \bar{D}\bar{E} \\ &= (\bar{\nabla}_\partial \bar{\nabla}_\partial \bar{D} - 2T_\gamma \bar{\nabla}_\partial \bar{D} + \bar{R}_\gamma \bar{D})\bar{E} \\ &= 0 \ . \end{aligned}$$

Therefore Y is the Jacobi field along γ with initial values

$$Y(0) = 0$$

and

$$(\nabla_\partial Y)(0) = \nabla_\partial(\bar{D}\bar{E})(0) = \bar{\nabla}_\partial(\bar{D}\bar{E})(0) - T_\gamma(\bar{D}\bar{E})(0) = (\bar{\nabla}_\partial \bar{D})\bar{E}(0) = \bar{E}(0) \ .$$

Using this fact we have, at $\gamma(r)$,

$$A\bar{D}\bar{E} = AY = -\nabla_Y \xi = -\nabla_\partial Y = -\bar{\nabla}_\partial Y + T_\gamma Y = -(\bar{\nabla}_\partial \bar{D})\bar{E} + T_\gamma \bar{D}\bar{E} \ ,$$

and therefore

$$A\bar{D} = -\bar{\nabla}_\partial \bar{D} + T_\gamma \bar{D} \ .$$

Since $\bar{D}(r)$ is non-singular (otherwise $\gamma(r)$ would be a conjugate point along γ, in which case $G_p(r)$ could not be a hypersurface), we may multiply the last equation from the right with $\bar{D}^{-1}$, by which the assertion follows. □

We will apply this lemma now to deduce

Theorem 1 *Every generalized Heisenberg group is an $\mathfrak{SC}$-space.*

Proof. We produce the situation described in the preceding lemma with M as a generalized Heisenberg group N, $p = e$ and $\dot{\gamma}(0) = V + Y \in \mathfrak{n}$. Our aim is to show that the eigenvalue functions of $A(r)$, considered as functions of r, are odd functions. Since $-A(-r)$ is the shape operator of $G_e(r)$ at $\gamma(-r)$, this then proves the assertion. We shall frequently use the computations carried out in the proof of Theorem 3.7. Once again we consider the three cases.

(i) $\underline{Y = 0}$. For $U \in \ker \mathrm{ad}(V) \cap V^\perp$ we have

$$T_\gamma U = 0\ ,\ \bar{\nabla}_\theta U = 0\ ,\ \bar{D}(r)U = rU\ ,$$

whence

$$A(r)U = -\frac{1}{r}U\ .$$

Next, let $U \in \ker \mathrm{ad}(V)^\perp$ be a unit vector. Then $X := [U,V] \in \mathfrak{z}$ is a unit vector and we have $U = -J_X V$. Both U and X are $\bar{\nabla}_\theta$-parallel and with respect to U, X we have, with $\sigma(U) := \mathrm{span}\{U, X\}$,

$$\bar{D}(r)|\sigma(U) = \frac{r}{6}\begin{pmatrix} 6 & -3r \\ 3r & 6 - r^2 \end{pmatrix}$$

and

$$T_\gamma(r)|\sigma(U) = \frac{1}{2}\begin{pmatrix} 0 & -1 \\ 1 & 0 \end{pmatrix}.$$

Then we get

$$\begin{aligned}
(\bar{\nabla}_\theta \bar{D})(r)|\sigma(U) &= \frac{1}{2}\begin{pmatrix} 2 & -2r \\ 2r & 2 - r^2 \end{pmatrix}, \\
\bar{D}^{-1}(r)|\sigma(U) &= \frac{2}{r(12 + r^2)}\begin{pmatrix} 6 - r^2 & 3r \\ -3r & 6 \end{pmatrix}, \\
(\bar{\nabla}_\theta \bar{D})(r)|\sigma(U) \circ \bar{D}^{-1}(r)|\sigma(U) &= \frac{1}{r(12 + r^2)}\begin{pmatrix} 12 + 4r^2 & -6r \\ 6r + r^3 & 12 \end{pmatrix},
\end{aligned}$$

and therefore,

$$A(r)|\sigma(U) = -\frac{1}{2r(12 + r^2)}\begin{pmatrix} 24 + 8r^2 & r^3 \\ r^3 & 24 \end{pmatrix}.$$

It follows that $\mathrm{tr}\, A|\sigma(U)$ is an odd function of r and $\mathrm{tr}\, A^2|\sigma(U)$ is an even function of r. From the characteristic equation for the eigenvalues we conclude that the eigenfunctions of $A|\sigma(U)$ are odd. Summing up, we see that all eigenfunctions of A are odd functions.

(ii) $\underline{V = 0}$. Then $\dot{\gamma} = Y$ and $R'_\gamma = R'_Y = 0$. So we may apply the Lemma with $T_\gamma = 0$ (hence, $\bar{\nabla}_\theta = \nabla_\theta$ and consequently, we drop the bar in the following). For $X \in Y^\perp$ we have $\nabla_Y X = 0$ and $D(r)X = rX$, and therefore

$$A(r)X = -\frac{1}{r}X\ .$$

Next, let $U \in \mathfrak{v}$ be a unit vector. Then

$$\nabla_Y(\cos(t/2)U + \sin(t/2)J_Y U) = 0 ,$$

which means that

$$E_U(t) := \cos(t/2)U + \sin(t/2)J_Y U$$

is the ∇_θ-parallel vector field along γ with initial value $E_U(0) = U$. According to Theorem 3.7 we have

$$B_U(t) = 2\sin(t/2)E_U(t) ,$$

and hence

$$D(r)E_U(r) = 2\sin(r/2)E_U(r) .$$

This implies

$$A(r)|\mathfrak{v} = -\frac{1}{2}\cot(r/2)id_{\mathfrak{v}} .$$

We again conclude that the eigenvalue functions of A are odd functions.

(iii) $\underline{V \neq 0 \neq Y}$. It was proved in 3.7 that

$$E_1(t) := -\frac{|Y|}{|V|}V(t) + \frac{|V|}{|Y|}Y , \ E_2(t) := \frac{1}{|V||Y|}J_Y V(t)$$

is a $\bar{\nabla}_\theta$-parallel orthonormal frame field of

$$\tilde{\mathfrak{n}}_3 := \mathfrak{n}_3 \cap (V+Y)^\perp ,$$

and that, with respect to E_1, E_2,

$$T_\gamma|\tilde{\mathfrak{n}}_3 = \frac{1}{2}\begin{pmatrix} 0 & -1 \\ 1 & 0 \end{pmatrix}$$

and

$$\bar{D}(r)|\tilde{\mathfrak{n}}_3 = \frac{1}{|Y|^3}\begin{pmatrix} \sin(|Y|r) - |V|^2|Y|r & |Y|(\cos(|Y|r)-1) \\ |Y|(1-\cos(|Y|r)) & |Y|^2\sin(|Y|r) \end{pmatrix} .$$

From this we obtain at once

$$(\bar{\nabla}_\theta\bar{D})(r)|\tilde{\mathfrak{n}}_3 = \frac{1}{|Y|^2}\begin{pmatrix} \cos(|Y|r) - |V|^2 & -|Y|\sin(|Y|r) \\ |Y|\sin(|Y|r) & |Y|^2\cos(|Y|r) \end{pmatrix} ,$$

$$\det \bar{D}(r)|\tilde{\mathfrak{n}}_3 = \frac{1}{|Y|^3}\nu(r)$$

with

$$\nu(r) := \frac{2}{|Y|}(1-\cos(|Y|r)) - |V|^2 r\sin(|Y|r) ,$$

and further,

$$\bar{D}^{-1}(r)|\tilde{\mathfrak{n}}_3 = \frac{1}{\nu(r)}\begin{pmatrix} |Y|^2\sin(|Y|r) & |Y|(1-\cos(|Y|r)) \\ |Y|(\cos(|Y|r)-1) & \sin(|Y|r) - |V|^2|Y|r \end{pmatrix} .$$

Therefore,

$$A(r)|\tilde{\mathfrak{n}}_3 =$$
$$-\frac{1}{2\nu(r)}\begin{pmatrix} 2|Y|^2\sin(|Y|r) & 2|Y|(1-\cos(|Y|r))-\nu(r) \\ 2|Y|(1-\cos(|Y|r))-\nu(r) & 2\sin(|Y|r)-2|V|^2|Y|r\cos(|Y|r) \end{pmatrix}.$$

This shows that $\operatorname{tr} A|\tilde{\mathfrak{n}}_3$ is an odd function of r and $\operatorname{tr} A^2|\tilde{\mathfrak{n}}_3$ is an even function of r. Thus the two eigenvalue functions of $A|\tilde{\mathfrak{n}}_3$ are odd functions.

Next, let $U \in \mathfrak{p}$ be a unit vector. Then, as above, it follows that

$$E_U(t) := \cos(|Y|t/2)U + \frac{1}{|Y|}\sin(|Y|t/2)J_Y U$$

is the ∇_ϑ-parallel vector field along γ with initial value $E_U(0) = U$. According to 3.7 we have

$$B_U(t) = \frac{2}{|Y|}\sin(|Y|t/2)E_U(t)\ ,$$

from which we readily compute

$$A(r)E_U(r) = -\frac{|Y|}{2}\cot(|Y|r/2)E_U(r)\ .$$

As $\mathfrak{p}$ is invariant under ∇_ϑ-parallel translation, it follows that

$$A(r)|\mathfrak{p} = -\frac{|Y|}{2}\cot(|Y|r/2)id_{\mathfrak{p}}\ .$$

Although all the necessary material is available to write down explicitly the shape operator on $\mathfrak{q}$, we omit this tedious work and restrict to the features of A which are needed to prove the required result.

First we consider the case $\mu_j \notin \{0,-1\}$. Let $X \in L_j$ be a unit vector, $E_1,\dots,E_6$ and $\mathfrak{q}_j(X)$ as in 3.7. The space $\mathfrak{q}_j(X)$ is invariant under T_γ and $\bar{R}_\gamma$ and $E_1,\dots,E_6$ are $\bar{\nabla}_\vartheta$-parallel, which shows that $\bar{D}(r)$ and hence also $A(r)$ maps $\mathfrak{q}_j(X)$ into itself. According to 3.7 we have

$$\bar{D}|\mathfrak{q}_j(X) \approx \Omega := \begin{pmatrix} o & e & e & o & o & e \\ e & o & o & e & e & o \\ e & o & o & e & e & o \\ o & e & e & o & o & e \\ o & e & e & o & o & e \\ e & o & o & e & e & o \end{pmatrix},$$

which means that with respect to $E_1,\dots,E_6$ the tensor field $\bar{D}|\mathfrak{q}_j(X)$ is a matrix function Ω for which each coefficient function is even (e) or odd (o). Note that if a coefficient function is zero we may choose e or o in the way we like. Also from 3.7 (see the matrix P there) we know that

$$T_\gamma|\mathfrak{q}_j(X) \approx \Omega\ .$$

From this we also obtain

$$\begin{aligned}
\det \bar{D}|\mathfrak{q}_j(X) &\approx e\,,\\
\bar{\nabla}_\partial \bar{D}|\mathfrak{q}_j(X) &\approx o\Omega\,,\\
\bar{D}^{-1}|\mathfrak{q}_j(X) &\approx \Omega\,,\\
\bar{\nabla}_\partial \bar{D}\circ\bar{D}^{-1}|\mathfrak{q}_j(X) &\approx o\Omega^2\approx\Omega\,,
\end{aligned}$$

and therefore,

$$A|\mathfrak{q}_j(X)\approx\Omega\,.$$

A straightforward computation now gives

$$\begin{array}{lllll}
\operatorname{tr} A|\mathfrak{q}_j(X)\approx o & , & \operatorname{tr} A^2|\mathfrak{q}_j(X)\approx e & , & \operatorname{tr} A^3|\mathfrak{q}_j(X)\approx o \quad ,\\
\operatorname{tr} A^4|\mathfrak{q}_j(X)\approx e & , & \operatorname{tr} A^5|\mathfrak{q}_j(X)\approx o & ; & \operatorname{tr} A^6|\mathfrak{q}_j(X)\approx e \quad ,
\end{array}$$

from which we conclude that the eigenvalue functions of $A|\mathfrak{q}_j(X)$ are odd functions.

Further, if $\mu_0=0$ and $X\in L_0$ is a unit vector, we consider (as in 3.7) the vector fields E_1, E_3, E_5 and get

$$\bar{D}|\mathfrak{q}_0(X)\approx\Omega:=\begin{pmatrix} o & e & o\\ e & o & e\\ o & e & o\end{pmatrix}\quad,\quad T_\gamma|\mathfrak{q}_0(X)\approx\Omega\,,$$

which, by a straightforward computation, yields

$$A|\mathfrak{q}_0(X)\approx\Omega\,.$$

Thus

$$\operatorname{tr} A|\mathfrak{q}_0(X)\approx o\,,\ \operatorname{tr} A^2|\mathfrak{q}_0(X)\approx e\,,\ \operatorname{tr} A^3|\mathfrak{q}_0(X)\approx o\,,$$

which shows that the eigenvalue functions of $A|\mathfrak{q}_0(X)$ are odd.

Finally, if $\mu_k=-1$ and $X\in L_k$ is a unit vector, we consider (as in 3.7) the vector fields E_1, E_2, E_3, E_4 to get

$$\bar{D}|\mathfrak{q}_k(X)\approx\Omega:=\begin{pmatrix} o & e & e & o\\ e & o & o & e\\ e & o & o & e\\ o & e & e & o\end{pmatrix}\quad,\quad T_\gamma|\mathfrak{q}_k(X)\approx\Omega\,,$$

and, by a straightforward computation, we now obtain

$$A|\mathfrak{q}_k(X)\approx\Omega\,.$$

Hence,

$$\operatorname{tr} A|\mathfrak{q}_k(X)\approx o\,,\ \operatorname{tr} A^2|\mathfrak{q}_k(X)\approx e\,,\ \operatorname{tr} A^3|\mathfrak{q}_k(X)\approx o\,,\ \operatorname{tr} A^4|\mathfrak{q}_k(X)\approx e\,,$$

which implies that the eigenvalue functions of $A|\mathfrak{q}_k(X)$ are again odd functions of r.

As $T_{\gamma(r)}G_e(r)$ can be decomposed orthogonally into $\tilde{\mathfrak{n}}_3$, $\mathfrak{p}$ and suitable spaces $\mathfrak{q}_j(X)$, we conclude that all eigenvalue functions of $A(r)$ are odd functions of r. This proves the assertion also in the last case $V\neq 0\neq Y$. □

The preceding Theorem and Corollary 2 in 3.6 then implies

Theorem 2. *Every generalized Heisenberg group is a $\mathfrak{TC}$-space.*

As any $\mathfrak{TC}$- or $\mathfrak{SC}$-space is a D'Atri space we now have an alternative proof of Theorem 4 in 3.2 stating that every generalized Heisenberg group is a D'Atri space.

Contracting twice the Gauss equation of second order, and using the constancy of the spectrum of the Jacobi operator along a geodesic γ, we obtain that the scalar curvature of $G_e(r)$ at $\gamma(r)$ is given by

$$-\frac{1}{4}mn - 2\mathrm{tr}\, R_{V+Y} + (\mathrm{tr}\, A_{\gamma(r)})^2 - \mathrm{tr}\, A^2_{\gamma(r)} \,.$$

As a consequence of Theorem 1 we therefore get

Corollary *The scalar curvature of any geodesic sphere in a generalized Heisenberg group is the same at antipodal points.*

3.10 Metric tensor with respect to normal coordinates

Let (M,g) be a k-dimensional Riemannian manifold, $p \in M$, $e_1, \ldots, e_k$ an orthonormal basis of T_pM and $x_1, \ldots, x_k$ the induced normal coordinates on an open neighborhood of p. Further, let (g_{ij}) be the matrix-valued map defined by

$$g_{ij} := g\left(\frac{\partial}{\partial x_i}, \frac{\partial}{\partial x_j}\right) .$$

So, (g_{ij}) is the metric tensor of M with respect to the normal coordinates $x_1, \ldots, x_k$. An elementary, but crucial, observation is the following

Lemma *If $x_1, \ldots, x_k$ and $\tilde{x}_1, \ldots, \tilde{x}_k$ are two systems of normal coordinates centered at p, then (g_{ij}) and $(\tilde{g}_{ij})$ differ only by conjugation with an orthogonal transformation. In particular, the eigenvalues of (g_{ij}) are independent of the choice of the normal coordinates centered at p.*

It was proved by J.E. D'Atri [Dat] that the eigenvalues of (g_{ij}) are the same at antipodal points (with respect to the center of the normal coordinates) if M is a naturally reductive Riemannian homogeneous space. This result was generalized by O. Kowalski and the third author [KoVa4] to the class of Riemannian g.o. spaces. Furthermore, they proved in [KoVa2] that this property of the metric also holds on commutative spaces. Up to now it was an open problem whether this geometric property of the metric characterizes Riemannian g.o. spaces or not. The following theorem shows that this is not the case.

Theorem *On every generalized Heisenberg group the eigenvalues of the metric with respect to normal coordinates are the same at antipodal points (with respect to the center of the normal coordinates).*

Proof. Clearly, by the homogeneity of a generalized Heisenberg group N, it suffices to study the metric with respect to normal coordinates centered at the identity e. Let $\gamma : \mathbb{R} \to N$ be a geodesic in M parametrized by arc length and with $\gamma(0) = e$. We define $V + Y := \dot{\gamma}(0)$, choose an orthonormal basis $e_1, \ldots, e_{n+m}$ of $T_e N \cong \mathfrak{n}$ such that $e_1 = V + Y$, and denote the resulting normal coordinates by $x_1, \ldots, x_{n+m}$. Let B_{e_i} denote the Jacobi field along γ with initial values $B_{e_i}(0) = 0$ and $B'_{e_i}(0) = e_i$. From the definition of normal coordinates it follows that

$$\dot{\gamma}(r) = \frac{\partial}{\partial x_1}(\gamma(r))$$

and

$$B_{e_i}(r) = r\frac{\partial}{\partial x_i}(\gamma(r)) \ , \ i = 2, \ldots, n+m \ .$$

Now let $\bar{\nabla}_\theta$ and $\bar{D}$ be as in Lemma 3.9. Further, let $\bar{E}_i$ be the $\bar{\nabla}_\theta$-parallel vector field along γ with $\bar{E}_i(0) = e_i$. According to Lemma 3.7 we then have

$$B_{e_i} = \bar{D}\bar{E}_i \ , \ i = 2, \ldots, n+m \ .$$

Thus, for $i, j = 2, \ldots, n+m$, we obtain

$$\begin{aligned} g_{ij}(\gamma(r)) &= g\left(\frac{\partial}{\partial x_i}, \frac{\partial}{\partial x_j}\right)(\gamma(r)) \\ &= \frac{1}{r^2} g(\bar{D}\bar{E}_i, \bar{D}\bar{E}_j)(r) \\ &= \frac{1}{r^2} g(\bar{D}^T\bar{D}\bar{E}_i, \bar{E}_j)(r) \ , \end{aligned}$$

and, as a consequence of the Gauss Lemma,

$$g_{1j}(\gamma(r)) = 0 \ .$$

As γ is parametrized by arc length, we also have

$$g_{11}(\gamma(r)) = g(\dot{\gamma}, \dot{\gamma})(r) = 1 \ .$$

Therefore, the metric (g_{ij}) along γ with respect to the normal coordinates is given by

$$(g_{ij}) = \frac{1}{r^2}\begin{pmatrix} r^2 & 0 & \cdots & 0 \\ 0 & & & \\ \vdots & & \bar{D}^T\bar{D} & \\ 0 & & & \end{pmatrix}$$

with respect to $\bar{E}_1, \ldots, \bar{E}_{n+m}$. So, in order to prove the assertion, it suffices to show that $\bar{D}^T\bar{D}(r)$ and $\bar{D}^T\bar{D}(-r)$ have the same eigenvalues for small $r \in \mathbb{R}_+$. The proof of this will be done along the same lines as the proof of Theorem 1 in 3.9, from which we also take the particular form of $\bar{D}$ in the three cases we study now.

(i) $\underline{Y = 0}$. If $U \in \ker \mathrm{ad}(V) \cap V^\perp$, then $B_U(r) = rU$ and hence

$$\bar{D}^T\bar{D}(r)U = r^2 U \ .$$

Next, let $U \in \ker \mathrm{ad}(V)^\perp$ be a unit vector. Then we have $U = -J_X V$ with $X := [U,V] \in \mathfrak{z}$. Both U and X are $\bar\nabla_0$-parallel, and with respect to U, X we obtain, with $\sigma(U) := \mathrm{span}\{U, X\}$,

$$\bar D|\sigma(U) \approx \begin{pmatrix} o & e \\ e & o \end{pmatrix} ,$$

which implies that $\mathrm{tr}\, \bar D^T \bar D|\sigma(U)$ and $\mathrm{tr}\, (\bar D^T \bar D)^2|\sigma(U)$ are even functions of r. From this we derive that the eigenvalues of $\bar D^T \bar D(r)|\sigma(U)$ and $\bar D^T \bar D(-r)|\sigma(U)$ are the same. As the orthogonal complement of $\mathbb{R}V$ in $\mathfrak{n}$ can be spanned by vectors and two-planes of the above form, the assertion in this case follows.

(ii) $\underline{V=0}$. In this case we have

$$\bar D(r)|Y^\perp = r\, id_{Y^\perp}$$

and

$$\bar D(r)|\mathfrak{v} = 2\sin(r/2) id_{\mathfrak{v}} ,$$

from which the assertion also follows.

(iii) $\underline{V \neq 0 \neq Y}$. We use the same notations as in the proof of Theorem 1 in 3.9. With respect to E_1, E_2 we have

$$\bar D|\tilde{\mathfrak{n}}_3 \approx \begin{pmatrix} o & e \\ e & o \end{pmatrix} ,$$

and hence

$$\bar D^T \bar D|\tilde{\mathfrak{n}}_3 \approx \begin{pmatrix} e & o \\ o & e \end{pmatrix} .$$

Therefore,

$$\mathrm{tr}\, \bar D^T \bar D|\tilde{\mathfrak{n}}_3 \approx e \ , \ \mathrm{tr}\, (\bar D^T \bar D)^2|\tilde{\mathfrak{n}}_3 \approx e \ ,$$

which shows that the eigenvalues of $\bar D^T \bar D|\tilde{\mathfrak{n}}_3$ are even functions of r.

Next, as

$$\bar D(r)|\mathfrak{p} = \frac{2}{|Y|} \sin(|Y|r/2) id_{\mathfrak{p}} ,$$

the eigenvalues of $\bar D^T \bar D|\mathfrak{p}$ are even functions of r.

If $\mu_j \notin \{0, -1\}$ and $X \in L_j$ is a unit vector, then

$$\bar D|\mathfrak{q}_j(X) \approx \begin{pmatrix} o & e & e & o & o & e \\ e & o & o & e & e & o \\ e & o & o & e & e & o \\ o & e & e & o & o & e \\ o & e & e & o & o & e \\ e & o & o & e & e & o \end{pmatrix} ,$$

with respect to $E_1, \ldots, E_6$, and hence

$$\bar{D}^T \bar{D}|\mathfrak{q}_j(X) \approx \begin{pmatrix} e & o & o & e & e & o \\ o & e & e & o & o & e \\ o & e & e & o & o & e \\ e & o & o & e & e & o \\ e & o & o & e & e & o \\ o & e & e & o & o & e \end{pmatrix} .$$

This implies

$$\begin{array}{lll} \operatorname{tr} \bar{D}^T \bar{D}|\mathfrak{q}_j(X) \approx e \ , & \operatorname{tr} (\bar{D}^T \bar{D})^2|\mathfrak{q}_j(X) \approx e \ , & \operatorname{tr} (\bar{D}^T \bar{D})^3|\mathfrak{q}_j(X) \approx e \ , \\ \operatorname{tr} (\bar{D}^T \bar{D})^4|\mathfrak{q}_j(X) \approx e \ , & \operatorname{tr} (\bar{D}^T \bar{D})^5|\mathfrak{q}_j(X) \approx e \ , & \operatorname{tr} (\bar{D}^T \bar{D})^6|\mathfrak{q}_j(X) \approx e \ , \end{array}$$

from which we conclude that the eigenvalue functions of $\bar{D}^T \bar{D}|\mathfrak{q}_j(X)$ are even functions.

If $\mu_0 = 0$ and $X \in L_0$ is a unit vector, then

$$\bar{D}|\mathfrak{q}_0(X) \approx \begin{pmatrix} o & e & o \\ e & o & e \\ o & e & o \end{pmatrix}$$

with respect to E_1, E_3, E_5. This implies

$$\bar{D}^T \bar{D}|\mathfrak{q}_0(X) \approx \begin{pmatrix} e & o & e \\ o & e & o \\ e & o & e \end{pmatrix} .$$

As above we conclude that the eigenvalue functions of $\bar{D}^T \bar{D}|\mathfrak{q}_0(X)$ are even functions.

If $\mu_k = -1$ and $X \in L_k$ is a unit vector, we have

$$\bar{D}|\mathfrak{q}_k(X) \approx \begin{pmatrix} o & e & e & o \\ e & o & o & e \\ e & o & o & e \\ o & e & e & o \end{pmatrix}$$

with respect to E_1, E_2, E_3, E_4, and therefore,

$$\bar{D}^T \bar{D}|\mathfrak{q}_k(X) \approx \begin{pmatrix} e & o & o & e \\ o & e & e & o \\ o & e & e & o \\ e & o & o & e \end{pmatrix} .$$

Also here we conclude easily that the eigenvalue functions of $\bar{D}^T \bar{D}|\mathfrak{q}_k(X)$ are even functions.

As the orthogonal complement of $V + Y$ in $\mathfrak{n}$ can be decomposed orthogonally into $\tilde{\mathfrak{n}}_3$, $\mathfrak{p}$ and suitable spaces $\mathfrak{q}_j(X)$, we obtain that all eigenvalue functions of $\bar{D}^T \bar{D}$ are even functions of r. This proves the assertion also in the case $V \neq 0 \neq Y$. □

Note that the property for the eigenvalues of the metric tensor with respect to normal coordinates provides, once again, a proof for the D'Atri property of generalized Heisenberg groups.

Chapter 4

Damek-Ricci spaces

In Section 1 we present basic material about the Damek-Ricci spaces. Most of the contents of this section is due to [Bog], [CoDoKoRi], [Dam1] and [Dam2]. The study of the geometry of certain left-invariant distributions in 4.1.10 and the proof for the (non-)existence of invariant nearly Kähler structures on Damek-Ricci spaces in 4.1.13 are new. In Section 2 we determine the complete spectrum of the Jacobi operator and the corresponding eigenspaces. The spectrum has also been written down in [Sza2] without providing a proof. We finish Section 2 with a detailed discussion of the sectional curvature. In Section 3 we provide two alternative proofs for the fact that a Damek-Ricci space is a $\mathfrak{C}$-space if and only if it is a symmetric space. The explicit knowledge of the eigenspaces of the Jacobi operator enables us also to prove that a Damek-Ricci space is a $\mathfrak{P}$-space if and only if it is symmetric. In Section 4 we provide an alternative proof for the harmonicity of Damek-Ricci spaces by showing that the distance function to the identity is isoparametric. In Section 5 we discuss several consequences of the results of the previous sections. Most of them are new geometrical characterizations of the symmetric manifolds among the Damek-Ricci spaces.

4.1 Basic concepts

4.1.1 Definition

Let $\mathfrak{n}$ be a generalized Heisenberg algebra, $\mathfrak{a}$ a one-dimensional real vector space and A a non-zero vector in $\mathfrak{a}$. We denote the inner product and the Lie bracket on $\mathfrak{n}$ by $<.,.>_{\mathfrak{n}}$ and $[.,.]_{\mathfrak{n}}$, respectively, and define a new vector space

$$\mathfrak{s} := \mathfrak{n} \oplus \mathfrak{a}$$

as the direct sum of $\mathfrak{n}$ and $\mathfrak{a}$. Each vector in $\mathfrak{s}$ can be written in a unique way in the form $V+Y+sA$ with some $V \in \mathfrak{v}$, $Y \in \mathfrak{z}$ and $s \in \mathbb{R}$. Throughout this chapter vectors in $\mathfrak{s}$ will be written in this form, and we always use the symbols U, V, W for vectors in $\mathfrak{v}$, X, Y, Z for vectors in $\mathfrak{z}$ and r, s, t for real numbers.

We now define an inner product $<.,.>$ and a Lie bracket $[.,.]$ on $\mathfrak{s}$ by

$$<U+X+rA, V+Y+sA> := <U+X, V+Y>_{\mathfrak{n}} + rs$$

and

$$[U+X+rA, V+Y+sA] := [U,V]_{\mathfrak{n}} + \frac{1}{2}rV - \frac{1}{2}sU + rY - sX .$$

In this way $\mathfrak{s}$ becomes a Lie algebra with an inner product. The attached simply connected Lie group, equipped with the induced left-invariant metric, is denoted by S and is called a *Damek-Ricci space.*

4.1.2 Classification and idea for construction

Of course, the classification of all Damek-Ricci spaces follows from the one of generalized Heisenberg groups established in 3.1.2. The dimensions of the Damek-Ricci spaces can be determined from the tables in 3.1.2 by just adding one dimension. Also here the case when m, the dimension of $\mathfrak{z}$, is congruent to $3(\mathrm{mod}\, 4)$ is of particular interest. In such a situation we define

$$\mathfrak{s}(k_1,k_2) := \mathfrak{n}(k_1,k_2) \oplus \mathfrak{a} ,$$

where k_1, k_2 are non-negative integers, and denote the corresponding Damek-Ricci space by $S(k_1,k_2)$. It follows from 3.1.2 that $\mathfrak{s}(k_1,k_2)$ is isomorphic to $\mathfrak{s}(\tilde{k}_1,\tilde{k}_2)$ if and only if for both Lie algebras the dimension of $\mathfrak{z}$ is the same and $(\tilde{k}_1,\tilde{k}_2) \in \{(k_1,k_2),(k_2,k_1)\}$; this is also the precise condition in order that two Damek-Ricci spaces $S(k_1,k_2)$ and $S(\tilde{k}_1,\tilde{k}_2)$ are isometric to each other.

The idea for the construction of the Lie algebras $\mathfrak{s}$ stems from the following considerations. Let $m = 1$, that is, $\mathfrak{n}$ is a $(2k+1)$-dimensional Heisenberg algebra in the classical sense. Then $\mathfrak{n}$ is isomorphic to the nilpotent part in the Iwasawa decomposition of the Lie algebra of the isometry group of the complex hyperbolic space $\mathbb{C}H^{k+1}$. The solvable part in this Iwasawa decomposition is isomorphic to $\mathfrak{s} = \mathfrak{n} \oplus \mathfrak{a}$ with the Lie algebra structure as defined above. This shows that for $m = 1$ the Damek-Ricci spaces are isometric to the complex hyperbolic spaces. Similarily, if $m = 3$, the Damek-Ricci space $S(k,0) \cong S(0,k)$ is isometric to the quaternionic hyperbolic space $\mathbb{H}H^{k+1}$; and if $m = 7$, the Damek-Ricci space $S(1,0) \cong S(0,1)$ is isometric to the Cayley hyperbolic plane $\mathrm{Cay}H^2$. So the construction of the Damek-Ricci spaces arises from imitating the construction of the non-compact rank-one symmetric spaces of non-constant curvature from their Iwasawa decomposition. Note that the real hyperbolic space can be obtained via this procedure by starting with a zero-dimensional vector space $\mathfrak{z}$ and, therefore, an Abelian Lie algebra $\mathfrak{n} = \mathfrak{v}$.

4.1.3 Algebraic features and diffeomorphism type

The derived subalgebra $[\mathfrak{s},\mathfrak{s}]$ of $\mathfrak{s}$ is equal to the generalized Heisenberg algebra $\mathfrak{n}$ and hence nilpotent. This shows that $\mathfrak{s}$ is a solvable Lie algebra and, therefore,

Proposition 1 *Each Damek-Ricci space is a solvable Lie group.*

The definition of the Lie algebra structure on $\mathfrak{s}$ implies that, as a Lie algebra, $\mathfrak{s}$ is the semi-direct sum $\mathfrak{s} = \mathfrak{n}+_f\mathfrak{a}$ of $\mathfrak{n}$ and $\mathfrak{a}$ with regard to the $\mathbb{R}$-algebra homomorphism

$$f : \mathfrak{a} \to \mathrm{der}(\mathfrak{n}) \ , \ sA \mapsto (\mathfrak{n} \to \mathfrak{n} \ , \ V{+}Y \mapsto \frac{1}{2}sV{+}sY) \ .$$

Carrying this over to the group level means that S is a semi-direct product $S = N \times_F \mathbb{R}$ of the generalized Heisenberg group N attached to S and $\mathbb{R}$, where

$$F : \mathbb{R} \to \mathrm{Aut}(N) \ , \ s \mapsto (N \to N \ , \ \exp_{\mathfrak{n}}(V{+}Y) \mapsto \exp_{\mathfrak{n}}(e^{s/2}V{+}e^{s}Y)) \ ,$$

$\exp_{\mathfrak{n}}$ is the Lie exponential map of N and $\mathbb{R}$ is considered in the canonical way as the simply connected Lie group attached to $\mathfrak{a}$. Writing this down explicitly, the group structure on $S = N \times_F \mathbb{R}$ is given by

$$\begin{aligned}&(\exp_{\mathfrak{n}}(U{+}X), r) \cdot (\exp_{\mathfrak{n}}(V{+}Y), s) \\ &= \left(\exp_{\mathfrak{n}}\left(U{+}e^{r/2}V{+}X{+}e^{r}Y{+}\frac{1}{2}e^{r/2}[U,V]\right), r{+}s\right) \ .\end{aligned}$$

Since $\exp_{\mathfrak{n}} : \mathfrak{n} \to N$ is a diffeomorphism, the map

$$\exp_{\mathfrak{n}} \times \exp_{\mathfrak{a}} : \mathfrak{n} \oplus \mathfrak{a} \to N \times \mathbb{R} \ , \ V{+}Y{+}sA \mapsto (\exp_{\mathfrak{n}}(V{+}Y), s)$$

is a diffeomorphism. Therefore the Lie group S is often identified with $\mathfrak{s}$ equipped with the group structure obtained from S via this diffeomorphism. We will not use this identification here. A consequence of the preceding consideration is

Proposition 2 *S is diffeomorphic to $\mathbb{R}^{m+n+1}$.*

We finally mention that the endomorphisms J_Z play also an important role for Damek-Ricci spaces, and their algebraic features stated in 3.1.3 will also be used frequently in this chapter without referring to them explicitly.

4.1.4 Lie exponential map

Let $V{+}Y{+}sA \in \mathfrak{s}$ be arbitrary. We want to compute now the corresponding one-parameter subgroup of S, that is, the integral curve $\alpha : \mathbb{R} \to S$ of $V{+}Y{+}sA$ with $\alpha(0) = e$. According to 4.1.3 we may write

$$\alpha(t) = (\exp_{\mathfrak{n}}(V(t) + Y(t)), s(t))$$

with some functions $V : \mathbb{R} \to \mathfrak{v}$, $Y : \mathbb{R} \to \mathfrak{z}$ and $s : \mathbb{R} \to \mathbb{R}$. Then the condition

$$\alpha(t + \tilde{t}) = \alpha(t) \cdot \alpha(\tilde{t})$$

leads to

$$\begin{aligned}&(\exp_{\mathfrak{n}}(V(t+\tilde{t}) + Y(t+\tilde{t})), s(t+\tilde{t})) \ = \\ &\left(\exp_{\mathfrak{n}}\left(V(t) + e^{s(t)/2}V(\tilde{t}) + Y(t) + e^{s(t)}Y(\tilde{t}) + \frac{1}{2}e^{s(t)/2}[V(t), V(\tilde{t})]\right), s(t) + s(\tilde{t})\right) \ .\end{aligned}$$

First, the second component gives

$$s(t+\tilde{t}) = s(t) + s(\tilde{t}) ,$$

and from

$$s(0) = 0 \ , \ s'(0) = s$$

we then get

$$s(t) = st \ .$$

As $\exp_{\mathfrak{n}}$ is a diffeomorphism, the first component then provides the conditions

$$\begin{aligned} V(t+\tilde{t}) &= V(t) + e^{st/2}V(\tilde{t}) , \\ Y(t+\tilde{t}) &= Y(t) + e^{st}Y(\tilde{t}) + \frac{1}{2}e^{st/2}[V(t), V(\tilde{t})] \ . \end{aligned}$$

Further, we have the initial conditions

$$V(0) = 0 \ , \ Y(0) = 0 \ ,$$

and, as the differential of $\exp_{\mathfrak{n}}$ at $0 \in \mathfrak{n}$ is just the identity transformation of $T_0\mathfrak{n} \cong \mathfrak{n}$,

$$V'(0) = V \ , \ Y'(0) = Y \ .$$

It can easily be checked that

$$V(t) := \begin{cases} tV & , \text{ if } s = 0 \\ \frac{2}{s}(e^{st/2} - 1)V & , \text{ if } s \neq 0 , \end{cases}$$

and

$$Y(t) := \begin{cases} tY & , \text{ if } s = 0 \\ \frac{1}{s}(e^{st} - 1)Y & , \text{ if } s \neq 0 , \end{cases}$$

are the solutions of these functional equations. Thus we summarize

Proposition *The one-parameter subgroup of $S = N \times_F \mathbb{R}$ induced by the vector $V+Y+sA \in \mathfrak{s}$ is*

$$\begin{aligned} &t \mapsto (\exp_{\mathfrak{n}}(t(V+Y)), 0) &, \text{ if } s = 0 , \\ &t \mapsto \left(\exp_{\mathfrak{n}}\left(\frac{2}{s}(e^{st/2} - 1)V + \frac{1}{s}(e^{st} - 1)Y\right), st\right) &, \text{ if } s \neq 0 . \end{aligned}$$

In particular, the Lie exponential map $\exp_{\mathfrak{s}} : \mathfrak{s} \to S$ satisfies

$$\exp_{\mathfrak{s}}(V+Y+sA) = \begin{cases} (\exp_{\mathfrak{n}}(V+Y), 0) & , \text{ if } s = 0 \\ \left(\exp_{\mathfrak{n}}\left(\frac{2}{s}(e^{s/2} - 1)V + \frac{1}{s}(e^{s} - 1)Y\right), s\right) & , \text{ if } s \neq 0 . \end{cases}$$

Using the fact that the Lie exponential map $\exp_{\mathfrak{n}}$ of N is a diffeomorphism, a simple calculation implies

Corollary *The Lie exponential map $\exp_{\mathfrak{s}} : \mathfrak{s} \to S$ of S is a diffeomorphism.*

The Lie exponential map on Damek-Ricci spaces has also been derived by J. Boggino [Bog]. Notice that he used another model of S, which is why his formula is not the same as ours.

4.1.5 Some global coordinates

We now introduce some global coordinates on S. Let $V_1, \ldots, V_n, Y_1, \ldots, Y_m, A$ be an orthonormal basis of the Lie algebra $\mathfrak{s}$ and $\tilde{v}_1, \ldots, \tilde{v}_n, \tilde{y}_1, \ldots, \tilde{y}_m, \tilde{\lambda}$ the corresponding coordinate functions on $\mathfrak{s}$. Using the diffeomorphism $\exp_{\mathfrak{n}} \times \exp_{\mathfrak{a}}$ we obtain global coordinates $v_1, \ldots, v_n, y_1, \ldots, y_m, \lambda$ on S via the relation

$$(v_1, \ldots, v_n, y_1, \ldots, y_m, \lambda) \circ (\exp_{\mathfrak{n}} \times \exp_{\mathfrak{a}}) = (\tilde{v}_1, \ldots, \tilde{v}_n, \tilde{y}_1, \ldots, \tilde{y}_m, \tilde{\lambda}) .$$

Lemma *We have*

$$\begin{aligned} V_i &= e^{\lambda/2}\frac{\partial}{\partial v_i} - \frac{1}{2}e^{\lambda/2}\sum_{j,k} A_{ij}^k v_j \frac{\partial}{\partial y_k} , \\ Y_i &= e^{\lambda}\frac{\partial}{\partial y_i} , \\ A &= \frac{\partial}{\partial \lambda} , \end{aligned}$$

where

$$A_{ij}^k := \langle [V_i, V_j], Y_k \rangle .$$

Proof. Let $p := (\exp_{\mathfrak{n}}(U + X), r) \in S = N \times_F \mathbb{R}$ be arbitrary and L_p the left translation on S by p. Then, using Proposition 4.1.4, we have

$$\begin{aligned} V_i(p) &= L_{p*e} V_i(e) \\ &= L_{p*e}\left(\frac{\partial}{\partial t}(t \mapsto (\exp_{\mathfrak{n}}(tV_i), 0))(0)\right) \\ &= \frac{\partial}{\partial t}(t \mapsto L_p(\exp_{\mathfrak{n}}(tV_i), 0))(0) \\ &= \dot{\alpha}_i(0) \end{aligned}$$

with

$$\alpha_i(t) = \left(\exp_{\mathfrak{n}}\left(U + e^{r/2}tV_i + X + \frac{1}{2}e^{r/2}t[U, V_i]\right), r\right) .$$

We have

$$\begin{aligned} (v_k \circ \alpha_i)(t) &= \tilde{v}_k(U) + \delta_{ik}e^{r/2}t , \\ (y_k \circ \alpha_i)(t) &= \tilde{y}_k(X) + \frac{1}{2}e^{r/2}t\langle [U, V_i], Y_k \rangle \\ &= \tilde{y}_k(X) - \frac{1}{2}e^{r/2}t\langle J_{Y_k}V_i, U \rangle \\ &= \tilde{y}_k(X) - \frac{1}{2}e^{r/2}t\sum_j \langle J_{Y_k}V_i, V_j \rangle\langle V_j, U \rangle \\ &= \tilde{y}_k(X) - \frac{1}{2}e^{r/2}t\sum_j A_{ij}^k \tilde{v}_j(U) , \\ (\lambda \circ \alpha_i)(t) &= r . \end{aligned}$$

Thus,

$$\begin{aligned}(v_k \circ \alpha_i)'(0) &= \delta_{ik} e^{r/2} , \\ (y_k \circ \alpha_i)'(0) &= -\frac{1}{2} e^{r/2} \sum_j A^k_{ij} \tilde{v}_j(U) , \\ (\lambda \circ \alpha_i)'(0) &= 0 ,\end{aligned}$$

and therefore,

$$\begin{aligned}V_i(p) &= \dot{\alpha}_i(0) \\ &= \sum_k (v_k \circ \alpha_i)'(0) \frac{\partial}{\partial v_k}(p) + \sum_k (y_k \circ \alpha_i)'(0) \frac{\partial}{\partial y_k}(p) + (\lambda \circ \alpha_i)'(0) \frac{\partial}{\partial \lambda}(p) \\ &= e^{\lambda(p)/2} \frac{\partial}{\partial v_i}(p) - \frac{1}{2} e^{\lambda(p)/2} \sum_{j,k} A^k_{ij} v_j(p) \frac{\partial}{\partial y_k}(p) .\end{aligned}$$

Similarily,

$$\begin{aligned}Y_i(p) &= L_{p*e} Y_i(e) \\ &= L_{p*e} \left(\frac{\partial}{\partial t}(t \mapsto (\exp_{\mathfrak{n}}(tY_i), 0))(0) \right) \\ &= \frac{\partial}{\partial t}(t \mapsto L_p(\exp_{\mathfrak{n}}(tY_i), 0))(0) \\ &= \dot{\beta}_i(0)\end{aligned}$$

with

$$\beta_i(t) = (\exp_{\mathfrak{n}}(U + X + e^r t Y_i), r) .$$

Here we have

$$(v_k \circ \beta_i)(t) = \tilde{v}_k(U) , \ (y_k \circ \beta_i)(t) = \tilde{y}_k(X) + \delta_{ik} e^r t , \ (\lambda \circ \beta_i)(t) = r ,$$

and hence

$$(v_k \circ \beta_i)'(0) = 0 , \ (y_k \circ \beta_i)'(0) = \delta_{ik} e^r , \ (\lambda \circ \beta_i)'(0) = 0 .$$

This implies

$$Y_i(p) = \dot{\beta}_i(0) = e^{\lambda(p)} \frac{\partial}{\partial y_i}(p) .$$

Finally,

$$\begin{aligned}A(p) &= L_{p*e} A(e) \\ &= L_{p*e} \left(\frac{\partial}{\partial t}(t \mapsto (\exp_{\mathfrak{n}}(0), t))(0) \right) \\ &= \frac{\partial}{\partial t}(t \mapsto L_p(\exp_{\mathfrak{n}}(0), t))(0) \\ &= \dot{c}(0)\end{aligned}$$

with

$$c(t) = (\exp_n(U+X), r+t) .$$

Here,

$$(v_k \circ c)(t) = \tilde{v}_k(U) , \ (y_k \circ c)(t) = \tilde{y}_k(X) , \ (\lambda \circ c)(t) = r+t ,$$

and hence

$$(v_k \circ c)'(0) = 0 , \ (y_k \circ c)'(0) = 0 , \ (\lambda \circ c)'(0) = 1 .$$

This implies

$$A(p) = \dot{c}(0) = \frac{\partial}{\partial \lambda}(p) .$$

Thus the Lemma is proved. □

We will use the Lemma also in the following converse form.

Corollary *We have*

$$\begin{aligned} \frac{\partial}{\partial v_i} &= e^{-\lambda/2} V_i + \frac{1}{2} e^{-\lambda} \sum_{j,k} A^k_{ij} v_j Y_k , \\ \frac{\partial}{\partial y_i} &= e^{-\lambda} Y_i , \\ \frac{\partial}{\partial \lambda} &= A . \end{aligned}$$

4.1.6 Levi Civita connection

Let ∇ be the Levi Civita connection of a Damek-Ricci space (S, g). Using the same method as for the generalized Heisenberg groups we obtain

$$\begin{aligned} &\nabla_{V+Y+sA}(U+X+rA) \\ &= -\frac{1}{2} J_X V - \frac{1}{2} J_Y U - \frac{1}{2} r V - \frac{1}{2} [U, V] - r Y + \frac{1}{2} \langle U, V \rangle A + \langle X, Y \rangle A . \end{aligned}$$

4.1.7 Curvature

Let R be the Riemannian curvature tensor, Q the $(1,1)$-Ricci tensor and τ the scalar curvature of a Damek-Ricci space (S, g). By a straightforward computation we get

$$\begin{aligned} &R(U+X+rA, V+Y+sA)(W+Z+tA) \\ &= \frac{1}{2} J_X J_Y W + \frac{1}{4} J_Z J_Y U - \frac{1}{4} J_Z J_X V + \frac{1}{2} J_{[U,V]} W - \frac{1}{4} J_{[V,W]} U + \frac{1}{4} J_{[U,W]} V \\ &\quad + \frac{1}{2} r J_Y W - \frac{1}{2} s J_X W - \frac{1}{4} s J_Z U + \frac{1}{4} t J_Y U + \frac{1}{4} r J_Z V - \frac{1}{4} t J_X V \\ &\quad + \frac{1}{2} \langle X, Y \rangle W - \frac{1}{4} (\langle V, W \rangle + st) U + \frac{1}{4} (\langle U, W \rangle + rt) V \end{aligned}$$

$$
\begin{aligned}
&-\frac{1}{2}[U,J_Z V]-\frac{1}{4}[U,J_Y W]+\frac{1}{4}[V,J_X W]+\frac{1}{2}t[U,V]+\frac{1}{4}s[U,W]-\frac{1}{4}r[V,W]\\
&-<V+Y+sA,W+Z+tA>X+<U+X+rA,W+Z+tA>Y\\
&+\frac{1}{2}<V,W>X-\frac{1}{2}<U,W>Y+\frac{1}{2}<U,V>Z\\
&+\Big\{-\frac{1}{2}<J_Z U,V>-\frac{1}{4}<J_Y U,W>+\frac{1}{4}<J_X V,W>\\
&\quad -r\left(\frac{1}{4}<V,W>+<Y,Z>\right)+s\left(\frac{1}{4}<U,W>+<X,Z>\right)\Big\}A\,,
\end{aligned}
$$

$$
\begin{aligned}
Q &= -\left(m+\frac{n}{4}\right) id_{\mathfrak{s}}\,,\\
\tau &= -(m+n+1)\left(m+\frac{n}{4}\right).
\end{aligned}
$$

In particular, the formula for the Ricci tensor shows that

Proposition 1 *Every Damek-Ricci space is an Einstein manifold.*

Further, let σ be a two-dimensional subspace of $T_eS \cong \mathfrak{s}$. Then there exist orthonormal vectors $U+X$ and $V+Y+sA$ in $\mathfrak{s}$ so that

$$\sigma = \text{span}\{U+X, V+Y+sA\}\,.$$

For the sectional $K(\sigma)$ of S with respect to σ we then have

$$K(\sigma) = -\frac{3}{4}|sX-[U,V]|^2-\frac{3}{4}<X,Y>^2-\frac{1}{4}(3|X|^2|Y|^2+6<J_X U, J_Y V>+1)\,.$$

A more detailed discussion of the sectional curvature can be found at the end of 4.2. In particular it will be proved there that

Proposition 2 *Every Damek-Ricci space is a Hadamard manifold.*

Recall that a *Hadamard manifold* is a complete and simply connected Riemannian manifold with non-positive sectional curvature. It is worthwhile to mention that a homogeneous manifold of non-positive curvature can always be represented as a solvable Lie group with a left-invariant metric (see for instance [Hei]). For a nice survey about the geometry of Riemannian manifolds of non-positive curvature see [Ebel]. A consequence of Proposition 2 is

Corollary *The exponential map* $\exp_e : T_eS \to S$ *of* S *at the identity* e *is a diffeomorphism.*

4.1.8 The Jacobi operator

Let $V+Y+sA \in \mathfrak{s}$ be a unit vector. For the Jacobi operator and its covariant derivative with respect to $V+Y+sA$ we then have

$$
\begin{aligned}
&R_{V+Y+sA}(U+X+rA) \\
&= \frac{3}{4}J_XJ_YV + \frac{3}{4}J_{[U,V]}V + \frac{3}{4}rJ_YV - \frac{3}{4}sJ_XV \\
&\quad -\frac{1}{4}U + \frac{3}{4}\langle X,Y\rangle V + \frac{1}{4}\langle U+X+rA, V+Y+sA\rangle V \\
&\quad -\frac{3}{4}[U,J_YV] + \frac{3}{4}s[U,V] - \left(1-\frac{3}{4}|V|^2\right)X + \langle U+X+rA, V+Y+sA\rangle Y \\
&\quad + \left\{\frac{3}{4}\langle U,J_YV\rangle - r\left(\frac{1}{4}|V|^2+|Y|^2\right) + s\left(\frac{1}{4}\langle U,V\rangle + \langle X,Y\rangle\right)\right\}A
\end{aligned}
$$

and

$$R'_{V+Y+sA}(U+X+rA) = \frac{3}{2}\left(J_{[U,V]}J_YV + J_{[U,J_YV]}V - \langle U,V\rangle J_YV - \langle U,J_YV\rangle V\right) .$$

4.1.9 Symmetry

The J^2-condition, which has been introduced in 3.1.3, can now be given a beautiful geometrical interpretation.

Theorem *A Damek-Ricci space S is a Riemannian symmetric space if and only if the attached generalized Heisenberg algebra $\mathfrak{n}$ satisfies the J^2-condition. More precisely, S is a Riemannian symmetric space if and only if*

(i) $m = 1$; then S is isometric to the complex hyperbolic space $\mathbb{C}H^{k+1}$, $2k = n$, with constant holomorphic sectional curvature -1; or

(ii) $m = 3$ and $\mathfrak{n} = \mathfrak{n}(k,0) \cong \mathfrak{n}(0,k)$; then S is isometric to the quaternionic hyperbolic space $\mathbb{H}H^{k+1}$ with constant quaternionic sectional curvature -1; or

(iii) $m = 7$ and $\mathfrak{n} = \mathfrak{n}(1,0) \cong \mathfrak{n}(0,1)$; then S is isometric the the Cayley hyperbolic plane $\mathrm{Cay}H^2$ with minimal sectional curvature -1.

Note that the simply connectedness of the Damek-Ricci spaces implies that if such a space were locally symmetric it would also be globally symmetric.

Proof. First suppose that S is a Riemannian symmetric space. Let $V+Y \in \mathfrak{n}$ be a unit vector with $V \neq 0 \neq Y$ and $K := K_{V,Y}$ as in 3.1.12. Further, let X be a non-zero vector in $Y^\perp$. Using the above expression for the covariant derivative of the Jacobi operator we obtain

$$0 = R'_{V+Y}J_XV = -\frac{3}{2}|V|^2(J_XJ_YV - J_{|Y|KX}V) .$$

This shows that $\mathfrak{n}$ satisfies the J^2-condition.

Conversely, suppose that $\mathfrak{n}$ satisfies the J^2-condition. Let $V+Y+sA \in \mathfrak{s}$ be a unit vector. Then we have

$$R'_{V+Y+sA}(X+rA) = 0 ,$$

$$\begin{aligned} R'_{V+Y+sA}V &= 0\,, \\ R'_{V+Y+sA}J_YV &= 0\,, \\ R'_{V+Y+sA}U &= 0 \text{ for all } U \in \ker \mathrm{ad}(V) \cap V^\perp\,, \\ R'_{V+Y+sA}J_XV &= -\frac{3}{2}|V|^2(J_XJ_YV - J_{|Y|K_X}V) = 0 \text{ for all } X \in Y^\perp\,. \end{aligned}$$

Since $\mathfrak{n}$ satisfies the J^2-condition, $\mathfrak{s}$ can be decomposed orthogonally into

$$\mathfrak{s} = (\mathfrak{z} \oplus \mathfrak{a}) \oplus \mathbb{R}V \oplus \mathbb{R}J_YV \oplus (\ker \mathrm{ad}(V) \cap V^\perp) \oplus J_{Y^\perp}V\,,$$

and we conclude that

$$R'_{V+Y+sA} = 0$$

for all unit vectors $V+Y+sA \in \mathfrak{s}$. This implies that $\nabla R = 0$ (see for instance Lemma 5.1 in [VaWi2]), that is, S is locally symmetric. As S is simply connected, it is also globally symmetric.

The assertion then follows from Theorem 3.1.3, the discussion at the end of 4.1.2 and the remarks about the values of sectional curvature at the end of 4.2. □

4.1.10 Integrability of certain subbundles

In this section we discuss the integrability of the subbundles

$$\mathfrak{v}\,,\ \mathfrak{z}\,,\ \mathfrak{a}\,,\ \mathfrak{v}\oplus\mathfrak{a}\,,\ \mathfrak{z}\oplus\mathfrak{a}\,,\ \mathfrak{n}$$

of TS obtained by left translation of the corresponding subspaces of $T_eS \cong \mathfrak{s}$ and the geometric structure of the induced foliations in case of integrability. Therefore we first recall the notions of Riemannian foliations, spherical submanifolds and isoparametric hypersurfaces.

Let (M,g) and B be Riemannian manifolds. A submersion $\pi : M \to B$ is called a *Riemannian submersion* if the differential of π preserves the length of all tangent vectors of M which are perpendicular to the fibers of π. Now, let L be a foliation of M and B the set of all leaves of L equipped with the quotient topology with respect to the canonical projection $\pi : M \to B$. Then L is called a *Riemannian foliation* if π can be made locally a Riemannian submersion. Geometrically this means that the leaves of L are locally equidistant (or parallel) to each other. If ∇ is the Levi Civita connection of M, then L is Riemannian if and only if for each vector field X tangent to L and all vector fields Y, Z orthogonal to L the equation

$$g(\nabla_YX, Z) + g(Y, \nabla_ZX) = 0$$

holds (see Theorem 5.19 in [Ton]). This analytic characterization of Riemannian foliations will be used in the proof of the subsequent Proposition.

A submanifold B of a Riemannian manifold M is called *spherical* (or an *extrinsic sphere*) if B is totally umbilical in M and the mean curvature vector field of B is parallel in the normal bundle of B.

A hypersurface B of a Riemannian manifold M is called *isoparametric* if its principal curvatures are constant.

Proposition *Concerning the integrability of the following left-invariant subbundles of TS we have*

(i) $\mathfrak{v}$ is not integrable;

(ii) $\mathfrak{z}$ is integrable and the induced foliation of S is Riemannian; each leaf is a spherical submanifold of S with mean curvature vector A and isometric to $\mathbb{R}^m$ with its standard Euclidean metric;

(iii) $\mathfrak{a}$ is integrable and the induced foliation of S is not Riemannian; each leaf is a totally geodesic submanifold of S and isometric to $\mathbb{R}$;

(iv) $\mathfrak{v} \oplus \mathfrak{a}$ is not integrable;

(v) $\mathfrak{z} \oplus \mathfrak{a}$ is integrable and the induced foliation of S is not Riemannian; each leaf is a totally geodesic submanifold of S and isometric to the real hyperbolic space $\mathbb{R}H^{m+1}$ of constant curvature -1;

(vi) $\mathfrak{v} \oplus \mathfrak{z} = \mathfrak{n}$ is integrable and the induced foliation of S is Riemannian; the foliation is the horosphere foliation of S by the horospheres centered at the point at infinity determined by the integral curves of A; each leaf is isometric to the corresponding generalized Heisenberg group N and an isoparametric hypersurface of S with two constant principal curvatures, $1/2$ and 1, and corresponding eigenspaces $\mathfrak{v}$ and $\mathfrak{z}$, respectively.

Proof. The statements (i) and (iv) follow from the fact that $\mathrm{ad}(V) : \mathfrak{v} \longrightarrow \mathfrak{z}$ is surjective for all non-zero $V \in \mathfrak{v}$.

For all $U+X, V+Y \in \mathfrak{n}$ we have

$$\nabla_{V+Y}(U+X) = -\frac{1}{2}J_X V - \frac{1}{2}J_Y U - \frac{1}{2}[U,V] + \frac{1}{2}<U,V>A + <X,Y>A \;,$$

and hence

$$[U+X, V+Y] \in \mathfrak{n} \,.$$

Therefore, $\mathfrak{n}$ is integrable. Moreover, as

$$<\nabla_A(U+X), A> + <A, \nabla_A(U+X)> = 0 \;,$$

$\mathfrak{n}$ induces a Riemannian foliation L. Now, according to 4.1.3, the multiplication on S restricted to $N \times_F \{0\}$ is just the multiplication on $N \times_F \{0\}$ induced from the one on N. Thus, the leaf of L through e is the hypersurface $N \times_F \{0\}$ and hence isometric to the generalized Heisenberg group N attached to S. By left-invariance of $\mathfrak{n}$, each leaf of $\mathfrak{n}$ is isometric to N. At each point A is a unit normal vector of the leaf through that point, and from

$$\nabla_V A = -\frac{1}{2}V \;\text{ and }\; \nabla_Y A = -Y$$

the statement about the principal curvatures follows. Finally, the statement about the horosphere foliation follows by applying Theorem 4.2 in [Wol2] to S. So statement (vi) is proved.

For all $X, Y \in \mathfrak{z}$ we have

$$\nabla_X Y = <X, Y> A \ .$$

This implies that $\mathfrak{z}$ is integrable, each leaf B of the induced foliation L is totally umbilical and A restricted to B is the mean curvature vector field of B. Since

$$\nabla_Y A = -Y$$

one sees that A is parallel in the normal bundle of B. Therefore each leaf of L is a spherical submanifold of S. Denote by R^B the Riemannian curvature tensor of B and by h the second fundamental form of B. Then

$$h(X, Y) = <X, Y> A \ ,$$

and the Gauss equation of second order implies, for $Z \in \mathfrak{z}$,

$$R^B(X,Y)Z = R(X,Y)Z + <Y, Z> X - <X, Z> Y = 0 \ ,$$

that is, B is a flat manifold. Now, let B be the leaf through the identity of S. As $\mathfrak{z} \subset \mathfrak{n}$, the leaf B is a submanifold of the generalized Heisenberg group N embedded in S as $N \times_F \{0\}$, the leaf of $\mathfrak{n}$ through e. According to 4.1.3 the multiplication on N coincides with the one on S restricted to $N \times_F \{0\}$. Thus we may apply Proposition 3.1.10 to $\mathfrak{z}|N$ and obtain that B is isometric to $\mathbb{R}^m$ with the standard Euclidean metric. By left-invariance of $\mathfrak{z}$ it follows that each leaf of L is isometric to $\mathbb{R}^m$. Further, for $X \in \mathfrak{z}$ and $V, W \in \mathfrak{v}$ we have

$$<\nabla_{V+sA} X, W+tA> + <V+sA, \nabla_{W+tA} X> = -\frac{1}{2}<J_X V, W> - \frac{1}{2}<V, J_X W> = 0 \ ,$$

which shows that L is a Riemannian foliation. Thus (ii) is proved.

As

$$\nabla_A A = 0 \ ,$$

each integral curve of A is a geodesic in S. Since S is complete, simply connected and of non-positive curvature it follows that each leaf of $\mathfrak{a}$ is a totally geodesic submanifold of S diffeomorphic (and hence isometric) to $\mathbb{R}$. Further, since

$$<\nabla_V A, V> + <V, \nabla_V A> = -|V|^2 \neq 0$$

for $V \neq 0$, the foliation cannot be Riemannian. This proves statement (iii).

Next, we have

$$\nabla_{Y+sA}(X+rA) = -rY + <X, Y> A \in \mathfrak{z} \oplus \mathfrak{a} \ .$$

This shows that $\mathfrak{z} \oplus \mathfrak{a}$ is integrable and the induced foliation L is totally geodesic. Let B be the leaf of $\mathfrak{z} \oplus \mathfrak{a}$ through e. Then the Gauss equation of second order implies

$$\begin{aligned} & R^B(X+rA, Y+sA)(Z+tA) \\ & \quad = -<Y+sA, Z+tA>(X+rA) + <X+rA, Z+tA>(Y+sA) \,, \end{aligned}$$

which is precisely the curvature tensor of a space form with curvature -1. According to Corollary 4.1.7 the exponential map $\exp_e$ of S at e is a diffeomorphism. This implies that $B = \exp_e(\mathfrak{z} \oplus \mathfrak{a})$ is diffeomorphic to $\mathbb{R}^{m+1}$ and hence isometric to the real hyperbolic space $\mathbb{R}H^{m+1}$ with constant sectional curvature -1. As

$$<\nabla_V A, V> + <V, \nabla_V A> = -|V|^2 \neq 0$$

for $V \neq 0$, L is not Riemannian. This proves statement (v). □

4.1.11 Geodesics

Let $V+Y+sA$ be a unit vector in $T_eS \cong \mathfrak{s}$ and $\gamma : \mathbb{R} \to S$ the geodesic in S with $\gamma(0) = e$ and $\dot\gamma(0) = V+Y+sA$. We define a subalgebra $\mathfrak{s}_4$ of $\mathfrak{s}$ by

$$\mathfrak{s}_4 := \mathrm{span}\{V, J_Y V, Y, A\} \,.$$

This algebra is one-dimensional if $V = 0 = Y$, two-dimensional if either $V = 0$ or $Y = 0$, and four-dimensional otherwise. A straightforward computation shows that the left-invariant subbundle $\mathfrak{s}_4$ of TS is autoparallel, that is, it is integrable and its leaves are totally geodesic. Clearly, γ lies in the leaf B_γ of the induced foliation of S through the identity e. Furthermore, B_γ is simply connected, for $B_\gamma = \exp_e(\mathfrak{s}_4)$ and $\exp_e$ is a diffeomorphism. If $V \neq 0 \neq Y$, the Lie algebra $\mathfrak{s}_4$ is the extension of the three-dimensional Heisenberg algebra as described in 4.1.1. Thus B_γ is isometric to the two-dimensional complex hyperbolic space $\mathbb{C}H^2$ with constant holomorphic sectional curvature -1. If $V = 0$ and $Y \neq 0$, then the Gauss equation of second order implies

$$R^{B_\gamma}\left(A, \frac{Y}{|Y|}\right)\frac{Y}{|Y|} = -A \quad \text{and} \quad R^{B_\gamma}\left(\frac{Y}{|Y|}, A\right)A = -\frac{Y}{|Y|} \,.$$

Thus B_γ is isometric to the one-dimensional complex hyperbolic space $\mathbb{C}H^1$ with constant sectional curvature -1. Furthermore, B_γ is the intersection of all totally geodesic $\mathbb{C}H^2$ in S containing e and at which Y and A are tangent. If $Y = 0$ and $V \neq 0$, then the Gauss equation of second order implies

$$R^{B_\gamma}\left(A, \frac{V}{|V|}\right)\frac{V}{|V|} = -\frac{1}{4}A \quad \text{and} \quad R^{B_\gamma}\left(\frac{V}{|V|}, A\right)A = -\frac{1}{4}\frac{V}{|V|} \,.$$

Thus B_γ is isometric to the two-dimensional real hyperbolic space $\mathbb{R}H^2$ with constant sectional curvature $-1/4$. Furthermore, B_γ is the intersection of all totally geodesic $\mathbb{C}H^2$ in S containing e and at which V and A are tangent. Finally, if

$V = 0 = Y$, γ is the integral curve of A and B_γ is the intersection of all totally geodesic $\mathbb{C}H^2$ in S containing e and at which A is tangent. The situation described above is very similar to the one in symmetric spaces, where the role of $\mathbb{C}H^2$ is taken over by the flats. Summing up we have shown

Proposition *Every geodesic in a Damek-Ricci space S lies in a suitable totally geodesically embedded complex hyperbolic space $\mathbb{C}H^2$ with constant holomorphic sectional curvature -1.*

As the geodesics in $\mathbb{C}H^2$ are well-known, the preceding Proposition enables one to write down explicitly the geodesics in Damek-Ricci spaces. The idea how to do this is due to [CoDoKoRi] and is as follows. As a model for the two-dimensional complex hyperbolic space we take the open unit ball

$$D := \{z \in \mathbb{C}^2 \mid |z| < 1\}$$

in $\mathbb{C}^2$ equipped with the Bergman metric

$$ds^2 = 4\frac{(1-|z|^2)dz \cdot d\bar{z} - \bar{z}dz \cdot zd\bar{z}}{(1-|z|^2)^2}$$

of constant holomorphic sectional curvature -1. Any geodesic $\alpha : \mathbb{R} \to D$ parametrized by arc length and with $\alpha(0) = 0$ is of the form

$$\alpha(t) = \theta(t)z$$

with some $z \in \partial D$ and

$$\theta(t) := \tanh(t/2)\ .$$

Next, consider the Cayley transform $C : D \to \tilde{D}$ from D onto the Siegel domain

$$\tilde{D} := \left\{(z_1, z_2) \in \mathbb{C}^2 \mid \Re(z_2) > \frac{1}{4}|z_1|^2\right\}$$

given by

$$C(z_1, z_2) = \left(\frac{2z_1}{1+z_2}, \frac{1-z_2}{1+z_2}\right)\ .$$

The map C is biholomorphic with inverse map

$$C^{-1}(z_1, z_2) = \left(\frac{z_1}{1+z_2}, \frac{1-z_2}{1+z_2}\right)\ .$$

Now equip $\tilde{D}$ with the Riemannian metric for which C becomes an isometry. This gives the Siegel domain model of the two-dimensional complex hyperbolic space equipped with the Bergman metric of constant holomorphic sectional curvature -1. We then deduce that

$$\beta := C \circ \alpha = \left(\frac{2\theta z_1}{1+\theta z_2}, \frac{1-\theta z_2}{1+\theta z_2}\right)$$

is the general form of a geodesic $\beta : \mathbb{R} \to \tilde{D}$ parametrized by arc length and with $\beta(0) = (0, 1)$.

Next, we fix unit vectors $\hat{V} \in \mathfrak{v}$ and $\hat{Y} \in \mathfrak{z}$ such that

$$V = |V|\hat{V} \text{ and } Y = |Y|\hat{Y}$$

and denote by $\mathbb{C}H^2$ the two-dimensional totally geodesic complex hyperbolic subspace in S determined by $\text{span}\{\hat{V}, J_{\hat{Y}}\hat{V}, \hat{Y}, A\}$. Note that $\mathbb{C}H^2 = B_\gamma$ if $V \neq 0 \neq Y$, and B_γ is strictly contained in $\mathbb{C}H^2$ otherwise. We define a bijection $\Phi : \tilde{D} \to \mathbb{C}H^2$ by

$$\Phi(z_1, z_2) := \left(\exp_{\mathfrak{n}}\left(\Re(z_1)\hat{V} + \Im(z_1)J_{\hat{Y}}\hat{V} + \Im(z_2)\hat{Y}\right), \ln\left(\Re(z_2) - \frac{1}{4}|z_1|^2\right)\right) .$$

Its inverse map is given by

$$\Phi^{-1}\left(\exp_{\mathfrak{n}}\left(a\hat{V} + bJ_{\hat{Y}}\hat{V} + c\hat{Y}\right), t\right) = \left(a + ib, e^t + \frac{1}{4}(a^2 + b^2) + ic\right) .$$

The Lie group $\mathbb{C}H^2$ acts simply transitively on $\tilde{D}$ by

$$\phi : \mathbb{C}H^2 \times \tilde{D} \to \tilde{D} , \ (g, p) \mapsto \Phi^{-1}(g \cdot \Phi(p)) ,$$

or explicitly, by

$$\begin{aligned} &\phi((\exp_{\mathfrak{n}}(a\hat{V} + bJ_{\hat{Y}}\hat{V} + c\hat{Y}), t), (z_1, z_2)) \\ &= \left(a + ib + e^{t/2}z_1, ic + \frac{1}{4}(a^2 + b^2) + \frac{1}{2}e^{t/2}(a - ib)z_1 + e^t z_2\right) . \end{aligned}$$

Thus $\mathbb{C}H^2$ acts on $\tilde{D}$ by biholomorphic transformations. As every biholomorphic transformation of $\tilde{D}$ is an isometry we conclude that $\mathbb{C}H^2$ acts on $\tilde{D}$ simply transitively by the isometries ϕ_g, $g \in \mathbb{C}H^2$. Next, since $\beta_1(0) = 0$ and $\beta_2(0) = 1$, we have

$$\Phi_{*(0,1)}\dot{\beta}(0) = \Re(\beta_1'(0))\hat{V} + \Im(\beta_1'(0))J_{\hat{Y}}\hat{V} + \Im(\beta_2'(0))\hat{Y} + \Re(\beta_2'(0))A .$$

From this we deduce that

$$\Phi_{*(0,1)} : T_{(0,1)}\tilde{D} \to T_e\mathbb{C}H^2$$

is a linear isometry. Since

$$\Phi \circ \phi_g = L_g \circ \Phi$$

for all $g \in \mathbb{C}H^2$, where L_g denotes left translation by g, and both L_g and ϕ_g are isometries, we conclude that Φ_{*z} is a linear isometry for each $z \in \tilde{D}$. Thus we have proved that Φ is an isometry. So the geodesic γ fixed at the beginning of this paragraph is given by $\gamma = \Phi \circ \beta$, where β is determined by the initial values

$$|V| = \beta_1'(0) = z_1 \text{ and } s + i|Y| = \beta_2'(0) = -z_2 .$$

As

$$\frac{1}{1 + \theta z_2} = \frac{1}{1 - s\theta - i|Y|\theta} = \frac{1}{\chi}(1 - s\theta + i|Y|\theta)$$

with

$$\chi := (1 - s\theta)^2 + |Y|^2\theta^2 ,$$

we eventually obtain

$$\begin{aligned}\gamma &= \Phi\left(\frac{2|V|\theta}{1-s\theta-i|Y|\theta},\frac{1+s\theta+i|Y|\theta}{1-s\theta-i|Y|\theta}\right)\\ &= \left(\exp_n\left(\frac{2\theta(1-s\theta)}{\chi}V+\frac{2\theta^2}{\chi}J_YV+\frac{2\theta}{\chi}Y\right),\ln\left(\frac{1-\theta^2}{\chi}\right)\right).\end{aligned}$$

So we have proved

Theorem 1 [CoDoKoRi] *Let $V+Y+sA \in \mathfrak{s}$ be a unit vector and $\gamma : \mathbb{R} \to S = N \times_F \mathbb{R}$ the geodesic in S with $\gamma(0) = e$ and $\dot\gamma(0) = V+Y+sA$. Then*

$$\gamma = \left(\exp_n\left(\frac{2\theta(1-s\theta)}{\chi}V+\frac{2\theta^2}{\chi}J_YV+\frac{2\theta}{\chi}Y\right),\ln\left(\frac{1-\theta^2}{\chi}\right)\right)$$

with

$$\theta(t) := \tanh(t/2) \quad \textit{and} \quad \chi := (1-s\theta)^2+|Y|^2\theta^2\ .$$

Note that in [CoDoKoRi] the second sign in the formula for the geodesics is not correct. Note also that in [CoDoKoRi] the model for S is a semi-direct product of N and $\mathbb{R}_+$ which can be obtained from ours by identifying $\mathbb{R}$ with $\mathbb{R}_+$ via the real exponential map.

We will now compute the tangent vector field $\dot\gamma$ of γ, where we still identify different tangent spaces along γ via left translation. Therefore, let $\hat V$, $\hat Y$ and $\mathbb{C}H^2$ as above. Further, let v, u, y, λ be the coordinate functions on $\mathbb{C}H^2$ induced by $\hat V, J_{\hat Y}\hat V, \hat Y, A$ (see 4.1.5). We define

$$h := \frac{1-\theta^2}{\chi}$$

and obtain, by using Corollary 4.1.5 and the preceding Theorem 1,

$$\begin{aligned}\dot\gamma &= (v\circ\gamma)'\frac{\partial}{\partial v}\circ\gamma+(u\circ\gamma)'\frac{\partial}{\partial u}\circ\gamma+(y\circ\gamma)'\frac{\partial}{\partial y}\circ\gamma+(\lambda\circ\gamma)'\frac{\partial}{\partial\lambda}\circ\gamma\\ &= \frac{1}{\sqrt{h}}(v\circ\gamma)'\hat V+\frac{1}{\sqrt{h}}(u\circ\gamma)'J_{\hat Y}\hat V\\ &\quad+\frac{1}{h}\left((y\circ\gamma)'+\frac{1}{2}(v\circ\gamma)'(u\circ\gamma)-\frac{1}{2}(u\circ\gamma)'(v\circ\gamma)\right)\hat Y\\ &\quad+(\lambda\circ\gamma)'A\ .\end{aligned}$$

By a straightforward computation we get

$$\begin{aligned}(v\circ\gamma)' &= |V|\frac{h}{\chi}((1-s\theta)^2-|Y|^2\theta^2)\ ,\\ (u\circ\gamma)' &= 2|V||Y|\frac{h}{\chi}\theta(1-s\theta)\ ,\\ (y\circ\gamma)' &= |Y|\frac{h}{\chi}(1-(|Y|^2+s^2)\theta^2)\ ,\\ (\lambda\circ\gamma)' &= (\ln h)'\ .\end{aligned}$$

Inserting these expressions into the general formula for $\dot\gamma$ we obtain

$$\dot\gamma = \frac{\sqrt{h}}{\chi}((1-s\theta)^2 - |Y|^2\theta^2)|V|\hat{V} + 2\frac{\sqrt{h}}{\chi}\theta(1-s\theta)|V||Y|J_{\hat{Y}}\hat{V} + h|Y|\hat{Y} + (\ln h)'A \ .$$

Therefore we have proved

Theorem 2 *Let $V{+}Y{+}sA \in \mathfrak{s}$ be a unit vector and $\gamma : \mathbb{R} \to S = N \times_F \mathbb{R}$ the geodesic in S with $\gamma(0) = e$ and $\dot\gamma(0) = V{+}Y{+}sA$. Then*

$$\dot\gamma = \frac{\sqrt{h}}{\chi}((1-s\theta)^2 - |Y|^2\theta^2)V + 2\frac{\sqrt{h}}{\chi}\theta(1-s\theta)J_Y V + hY + (\ln h)'A \ ,$$

where

$$\theta(t) := \tanh(t/2) \ , \ \chi := (1-s\theta)^2 + |Y|^2\theta^2 \text{ and } h := \frac{1-\theta^2}{\chi} \ .$$

In particular, when we decompose $\dot\gamma(t)$ into

$$\dot\gamma(t) = V(t) + Y(t) + A(t)$$

with respect to $\mathfrak{s} = \mathfrak{v} \oplus \mathfrak{z} \oplus \mathfrak{a}$, then

$$|V(t)|^2 = |V|^2 h(t) \ \text{ and } \ |Y(t)|^2 = |Y|^2 h^2(t) \ .$$

The last statement can be checked easily by using the explicit formula for $\dot\gamma$. In particular we see now that in the generic case $V \neq 0 \neq Y$ the length of the projections of $\dot\gamma$ onto $\mathfrak{v}$ and $\mathfrak{z}$ is not constant along γ.

4.1.12 Isometry group

The isometry groups of the symmetric Damek-Ricci spaces, that is, of $\mathbb{C}H^k$, $\mathbb{H}H^k$ ($k \geq 2$) and CayH^2, are well-known from the classical theory of symmetric spaces. If S is non-symmetric, the isometry group of S is as small as possible. Denote by $A(S)$ the group of automorphisms of S whose differential at e preserves the inner product on $T_eS \cong \mathfrak{s}$, and by $L(S)$ the group of left translations on S. Clearly, any map belonging to $A(S)$ or $L(S)$ is an isometry of S.

Theorem [Dam2] *If S is a non-symmetric Damek-Ricci space, the isometry group of S is the semi-direct product $L(S) \times_F A(S)$ with*

$$F : A(S) \to \text{Aut}(L(S)) \ , \ \phi \mapsto (L(S) \to L(S) \ , \ L_g \mapsto \phi L_g \phi^{-1} = L_{\phi(g)}) \ .$$

4.1.13 Nearly Kähler structures

Recall that a *nearly Kähler structure* on a Riemannian manifold (M,g) is a skew-symmetric tensor field J of type (1,1) on M satisfying $J^2 = -id_{TM}$ and

$$(\nabla_v J)v = 0 ,$$

or equivalently, by polarization,

$$(\nabla_v J)w + (\nabla_w J)v = 0$$

for all tangent vectors $v, w \in T_pM$ and all $p \in M$.

Suppose now that J is a nearly Kähler structure on a Damek-Ricci space S which is invariant under the group $L(S)$ of left translations on S. So each left translation is a holomorphic map of S with respect to J, and J maps left-invariant vector fields on S to left-invariant ones. We denote by $(\ )_{\mathfrak{v}}$ and $(\ )_{\mathfrak{z}}$ the projections onto $\mathfrak{v}$ and $\mathfrak{z}$, respectively. From

$$\begin{aligned} 0 &= -\nabla_A(JY) + J\nabla_A Y \\ &= -(\nabla_A J)Y \\ &= (\nabla_Y J)A \\ &= \nabla_Y(JA) - J\nabla_Y A \\ &= -\frac{1}{2}J_Y(JA)_{\mathfrak{v}} + \langle JA, Y\rangle A + JY \end{aligned}$$

we get the relations

$$\begin{aligned} 2(JY)_{\mathfrak{v}} &= J_Y(JA)_{\mathfrak{v}} , \\ (JY)_{\mathfrak{z}} &= 0 . \end{aligned}$$

Similarily, we obtain from

$$0 = -(\nabla_A J)V = (\nabla_V J)A = -\frac{1}{2}J_{(JA)_{\mathfrak{z}}}V - \frac{1}{2}[(JA)_{\mathfrak{v}}, V] + \frac{1}{2}\langle JA, V\rangle A + \frac{1}{2}JV$$

the relation

$$(JV)_{\mathfrak{z}} = [(JA)_{\mathfrak{v}}, V] .$$

Using these relations we obtain

$$-\langle JY, V\rangle = \langle JV, Y\rangle = \langle [(JA)_{\mathfrak{v}}, V], Y\rangle = \langle J_Y(JA)_{\mathfrak{v}}, V\rangle = 2\langle JY, V\rangle ,$$

and therefore also

$$(JY)_{\mathfrak{v}} = 0 .$$

We conclude that $J\mathfrak{z} \subset \mathfrak{a}$ and hence $\dim \mathfrak{z} = 1$, that is, S is a complex hyperbolic space. We summarize

Theorem *A Damek-Ricci space S admits a nearly Kähler structure which is invariant under the group $L(S)$ of left translations on S if and only if S is isometric to a complex hyperbolic space.*

Complex hyperbolic spaces are known to carry even an invariant Kähler structure, that is, an invariant nearly Kähler structure for which also $\nabla J = 0$ holds. That none of the Damek-Ricci spaces with $\dim \mathfrak{z} \geq 2$ carries a Kähler structure (which is invariant under $L(S)$) follows also from a more general theory. Indeed, since the scalar curvature of a Damek-Ricci space S is non-zero, any Kähler structure on S has to be invariant. Moreover, since S is a solvable Lie group it must be a homogeneous domain and finally, because it has non-positive curvature, S has to be symmetric. See [DaDo], [Dor], [DoNa], [Lic1], and [Lic2] for more details.

Recently an analogue of the above mentioned result by D'Atri and Dotti Miatello has been discovered in the framework of quaternionic geometry. In [Cor] it is proved that a (non-flat) real solvable Lie group endowed with an invariant quaternionic Kähler structure (called an *Alekseevskii space*) and of non-positive curvature is symmetric. This implies that *a non-symmetric Damek-Ricci space cannot be equipped with an invariant quaternionic Kähler structure.* About the classification of the Alekseevskii spaces, see [Ale], [Cor] and [dWVP].

4.2 Spectral properties of the Jacobi operator

In this section we compute the eigenvalues and the corresponding eigenspaces of the Jacobi operators of an arbitrary Damek-Ricci space S at the identity e. For a partial result see also [Sza2].

Theorem *Let $V+Y+sA$ be a unit vector in $\mathfrak{s}$.*

(i) $\underline{V=0, Y=0}$. *The eigenvalues and eigenspaces of R_{sA} are*

$$\begin{array}{rl} 0 & , \ \mathfrak{a}\ ; \\ -1/4 & , \ \mathfrak{v}\ ; \\ -1 & , \ \mathfrak{z}\ . \end{array}$$

(ii) $\underline{V=0, s=0}$. *The eigenvalues and eigenspaces of R_Y are*

$$\begin{array}{rl} 0 & , \ \mathbb{R}Y\ ; \\ -1/4 & , \ \mathfrak{v}\ ; \\ -1 & , \ Y^\perp \oplus \mathfrak{a}\ . \end{array}$$

(iii) $\underline{Y=0, s=0}$. *The eigenvalues and eigenspaces of R_V are*

$$\begin{array}{rl} 0 & , \ \mathbb{R}V\ ; \\ -1/4 & , \ (\ker \mathrm{ad}(V) \cap V^\perp) \oplus \mathfrak{z} \oplus \mathfrak{a}\ ; \\ -1 & , \ \ker \mathrm{ad}(V)^\perp\ . \end{array}$$

(iv) $\underline{V=0, Y\neq 0, s\neq 0}$. *The eigenvalues and eigenspaces of R_{Y+sA} are*

$$\begin{array}{rl} 0 & , \ \mathbb{R}(Y+sA)\ ; \\ -1/4 & , \ \mathfrak{v}\ ; \\ -1 & , \ (\mathfrak{z} \oplus \mathfrak{a}) \cap (Y+sA)^\perp\ . \end{array}$$

(v) $V \neq 0, Y = 0, s \neq 0$. *The eigenvalues and eigenspaces of* R_{V+sA} *are*

$$\begin{array}{rl} 0 \ , & \mathbb{R}(V+sA) \ ; \\ -1/4 \ , & (\ker \mathrm{ad}(V) \oplus \mathfrak{a}) \cap (V+sA)^{\perp} \oplus \{|V|^2 X - sJ_X V \mid X \in \mathfrak{z}\} \ ; \\ -1 \ , & \{sX + J_X V \mid X \in \mathfrak{z}\} \ . \end{array}$$

(vi) $V \neq 0 \neq Y$. *We decompose* $\mathfrak{s}$ *orthogonally into*

$$\mathfrak{s} = \mathfrak{s}_4 \oplus \mathfrak{p} \oplus \mathfrak{q} \ ,$$

where

$$\begin{aligned} \mathfrak{s}_4 &:= \mathrm{span}\{V, J_Y V, Y, A\} \ , \\ \mathfrak{p} &:= \ker \mathrm{ad}(V) \cap \ker \mathrm{ad}(J_Y V) \ , \\ \mathfrak{q} &:= \mathrm{span}\{Y^{\perp}, J_{Y^{\perp}} V, J_{Y^{\perp}} J_Y V\} \ . \end{aligned}$$

The spaces $\mathfrak{s}_4$, $\mathfrak{p}$ *and* $\mathfrak{q}$ *are invariant under* R_{V+Y+sA} *and we have*

(1) the eigenvalues and eigenspaces of $R_{V+Y+sA}|\mathfrak{s}_4$ *are*

$$\begin{array}{rl} 0 \ , & \mathbb{R}(V+Y+sA) \ ; \\ -1/4 \ , & \{(\alpha s + \beta |Y|^2) V + (\beta s - \alpha) J_Y V - \beta |V|^2 Y - \alpha |V|^2 A \mid \alpha, \beta \in \mathbb{R}\} \ ; \\ -1 \ , & \mathbb{R}(J_Y V + sY - |Y|^2 A) \ ; \end{array}$$

(2) (if $\mathfrak{p} \neq \{0\}$*)* $R_{V+Y+sA}|\mathfrak{p}$ *has only one eigenvalue, namely* $-1/4$;

(3) (if $\mathfrak{q} \neq \{0\}$*) We put* $K := K_{V,Y}$ *and decompose* $Y^{\perp}$ *orthogonally into*

$$Y^{\perp} = L_0 \oplus \ldots \oplus L_k \ ,$$

where

$$L_j := \ker(K^2 - \mu_j id_{Y^{\perp}}) \ , \quad (j = 0, \ldots, k)$$

and

$$0 \geq \mu_0 > \mu_1 > \ldots > \mu_k \geq -1$$

are the distinct eigenvalues of K^2. *It can easily be seen that*

$$X \in L_j \Longrightarrow KX \in L_j \quad (j = 0, \ldots, k) \ ,$$

whence $\dim L_j$ *is even provided that* $\mu_j \neq 0$. *We now define*

$$\begin{aligned} \mathfrak{q}_j &:= \mathrm{span}\{L_j, J_{L_j} V, J_{L_j} J_Y V\} \ , \ j = 0, \ldots, k, \ \mu_k \neq -1 \ , \\ \mathfrak{q}_k &:= \mathrm{span}\{L_k, J_{L_k} V\} \ , \ if \ \mu_k = -1 \ . \end{aligned}$$

Then

$$\mathfrak{q} = \mathfrak{q}_0 \oplus \ldots \oplus \mathfrak{q}_k \ , \ \dim \mathfrak{q}_j \equiv \begin{cases} 0 (\mathrm{mod}\, 3) & , \ if \ \mu_j = 0 \\ 0 (\mathrm{mod}\, 4) & , \ if \ \mu_j = -1 \\ 0 (\mathrm{mod}\, 6) & , \ otherwise \end{cases}$$

and each space $\mathfrak{q}_j$ *is invariant under* R_{V+Y+sA}. *Further, we have*

(A) (if $j = k$ and $\mu_k = -1$) the eigenvalues and eigenspaces of $R_{V+Y+sA}|\mathfrak{q}_k$ are

$$\begin{aligned} -1/4 &\ , \ \{|V|^2X + J_X(J_YV - sV) \mid X \in L_k\} \ ; \\ -1 &\ , \ \{(|V|^2 - 1)X + J_X(J_YV - sV) \mid X \in L_k\} \ . \end{aligned}$$

(B) (otherwise) $R_{V+Y+sA}|\mathfrak{q}_j$ has two or three distinct eigenvalues κ_i, $i = 1, 2, 3$, and

$$-1 < \kappa_1 \le -\frac{3}{4} \le \kappa_2 < -\frac{1}{4} < \kappa_3 \le 0 \ ,$$

where

$$\kappa_1 = \kappa_2 \Longleftrightarrow j = 0 \ , \ \mu_0 = 0 \ , \ s = 0 \ , \ |V|^2 = \frac{2}{3} \ .$$

The eigenvalues are the solutions of

$$(\kappa + 1)\left(\kappa + \frac{1}{4}\right)^2 = \frac{27}{64}|V|^4|Y|^2(1 + \mu_j)$$

and satisfy the relation

$$\kappa_1 + \kappa_2 + \kappa_3 = -\frac{3}{2} \ .$$

Furthermore, we have

(a) (if $j = 0$, $\mu_0 = 0$ and $s = 0$) the eigenvalues and eigenspaces of $R_{V+Y}|\mathfrak{q}_0$ are

$$\begin{aligned} -\frac{1}{4}(1 + 3|V|^2) &\ , \ J_{L_0}V \ ; \\ \frac{1}{8}(3|V|^2 - 5 \pm 3|Y|\sqrt{1 + 3|V|^2}) &\ , \ \{(4\kappa + 1)X + 3J_XJ_YV \mid X \in L_0\}; \end{aligned}$$

(b) (otherwise) the eigenspace of $R_{V+Y+sA}|\mathfrak{q}_j$ with respect to κ_i is

$$\begin{aligned} &\{(4\kappa_i + 1)(4\kappa_i + 1 + 3|V|^2)X + 3(4\kappa_i + 1 + 3|V|^2)J_X(J_YV - sV) \\ &-9|V|^2 J_{|Y|KX - sX}V \mid X \in L_j\} \ . \end{aligned}$$

Proof. As $R_{V+Y+sA}(V+Y+sA) = 0$, we will always assume that $U+X+rA \in \mathfrak{s}$ is orthogonal to $V+Y+sA$. Then we have

$$\begin{aligned} &R_{V+Y+sA}(U+X+rA) \\ = \ & \frac{3}{4}J_XJ_YV + \frac{3}{4}J_{[U,V]}V + \frac{3}{4}rJ_YV - \frac{3}{4}sJ_XV - \frac{1}{4}U + \frac{3}{4}\langle X, Y\rangle V \\ & -\frac{3}{4}[U, J_YV] + \frac{3}{4}s[U, V] - \left(1 - \frac{3}{4}|V|^2\right)X \\ & + \left\{\frac{3}{4}\langle U, J_YV\rangle - r\left(\frac{1}{4}|V|^2 + |Y|^2\right) + s\left(\frac{1}{4}\langle U, V\rangle + \langle X, Y\rangle\right)\right\}A \ . \end{aligned}$$

We now consider six cases.

(i) $\underline{V=0, Y=0}$. Then

$$R_{sA}(U+X) = -\frac{1}{4}U - X ,$$

and the assertion follows.

(ii) $\underline{V=0, s=0}$. Then

$$R_Y(U+X+rA) = -\frac{1}{4}U - X - rA ,$$

and the statement follows.

(iii) $\underline{Y=0, s=0}$. Then

$$R_V(U+X+rA) = \frac{3}{4}J_{[U,V]}V - \frac{1}{4}U - \frac{1}{4}X - \frac{1}{4}rA .$$

If $U+X+rA \in (\ker \mathrm{ad}(V) \cap V^\perp) \oplus \mathfrak{z} \oplus \mathfrak{a}$, then it is an eigenvector of R_V with corresponding eigenvalue $-1/4$. If $U \in \ker \mathrm{ad}(V)^\perp$, then there exists a vector $Z \in \mathfrak{z}$ so that $U = J_Z V$ and we get

$$R_V U = \frac{3}{4}J_{[J_Z V,V]}V - \frac{1}{4}U = -\frac{3}{4}J_Z V - \frac{1}{4}U = -U .$$

This proves the assertion.

(iv) $\underline{V=0, Y\neq 0, s\neq 0}$. Then $<X,Y> = -rs$, $|Y|^2+s^2=1$, and hence

$$R_{Y+sA}(U+X+rA) = -\frac{1}{4}U - X - rA ,$$

which proves the assertion.

(v) $\underline{V\neq 0, Y=0, s\neq 0}$. Since $<U,V> = -rs$ and $|V|^2+s^2=1$, we get

$$R_{V+sA}(U+X+rA) = \frac{3}{4}J_{[U,V]}V - \frac{3}{4}sJ_XV - \frac{1}{4}U + \frac{3}{4}s[U,V] - \frac{1}{4}(1+3s^2)X - \frac{1}{4}rA .$$

First suppose that $U+rA \in \ker \mathrm{ad}(V) \oplus \mathfrak{a}$. Then

$$R_{V+sA}(U+rA) = -\frac{1}{4}(U+rA) .$$

Next, if $0 \neq X \in \mathfrak{z}$, then

$$\begin{aligned} R_{V+sA}X &= -\frac{3}{4}sJ_XV - \frac{1}{4}(1+3s^2)X , \\ R_{V+sA}J_XV &= -\frac{3}{4}s|V|^2X - \left(1-\frac{3}{4}s^2\right)J_XV . \end{aligned}$$

Thus R_{V+sA} maps $\mathrm{span}\{X, J_XV\}$ into itself. Let κ be an eigenvalue of R_{V+sA} restricted to this 2-plane. Then there exist $\alpha, \beta \in \mathbb{R}$ so that $\alpha X + \beta J_XV$ is a corresponding eigenvector. As $\beta = 0$ is impossible (since X is not an eigenvector of R_{V+sA}), we may put $\beta = -3s/4$. Then the eigenvector equation

$$R_{V+sA}\left(\alpha X - \frac{3}{4}sJ_XV\right) = \kappa\alpha X - \frac{3}{4}\kappa sJ_XV$$

yields

$$\begin{aligned}\kappa\alpha &= -\frac{1}{4}(1+3s^2)\alpha + \frac{9}{16}s^2|V|^2 , \\ \kappa &= \alpha - 1 + \frac{3}{4}s^2 .\end{aligned}$$

From the second equation we get

$$\alpha = \kappa + 1 - \frac{3}{4}s^2 ,$$

and inserting this into the first equation then implies

$$(\kappa+1)\left(\kappa+\frac{1}{4}\right) = 0 .$$

Thus $\kappa \in \{-1/4, -1\}$. If $\kappa = -1/4$, then $\alpha = 3(1-s^2)/4 = 3|V|^2/4$; and for $\kappa = -1$ we get $\alpha = -3s^2/4$. From this the assertion finally follows.

(vi) $\underline{V \neq 0, Y \neq 0}$. Clearly, $R_{V+Y+sA}(V+Y+sA) = 0$. A straightforward computation shows that

$$R_{V+Y+sA}(J_Y V + sY - |Y|^2 A) = -(J_Y V + sY - |Y|^2 A)$$

and

$$\begin{aligned}&R_{V+Y+sA}((\alpha s + \beta|Y|^2)V + (\beta s - \alpha)J_Y V - \beta|V|^2 Y - \alpha|V|^2 A) \\ &= -\frac{1}{4}((\alpha s + \beta|Y|^2)V + (\beta s - \alpha)J_Y V - \beta|V|^2 Y - \alpha|V|^2 A) .\end{aligned}$$

Thus the statement about the eigenvalues and eigenspaces of R_{V+Y+sA} on $\mathfrak{s}_4$ is proved.

Next, for all $U \in \mathfrak{p}$ one easily derives

$$R_{V+Y+sA} U = -\frac{1}{4} U .$$

We continue by proving that each $\mathfrak{q}_j$ is invariant under R_{V+Y+sA}. For $X \in L_j$ with $\mu_j \neq -1$ we obtain

$$\begin{aligned}R_{V+Y+sA} X &= \frac{3}{4} J_X J_Y V - \frac{3}{4} s J_X V - \left(1 - \frac{3}{4}|V|^2\right) X , \\ R_{V+Y+sA} J_X V &= -\frac{1}{4}(1 + 3|V|^2) J_X V - \frac{3}{4}|V|^2 |Y| K X - \frac{3}{4} s |V|^2 X , \\ R_{V+Y+sA} J_X J_Y V &= -\frac{3}{4}|V|^2 |Y| J_{KX} V - \frac{1}{4} J_X J_Y V + \frac{3}{4}|V|^2 |Y|^2 X - \frac{3}{4} s |V|^2 |Y| K X .\end{aligned}$$

If $j = k$ and $\mu_k = -1$, then $J_X J_Y V = J_{|Y|KX} V$ and hence

$$\begin{aligned}R_{V+Y+sA} X &= \frac{3}{4} J_{|Y|KX} V - \frac{3}{4} s J_X V - \left(1 - \frac{3}{4}|V|^2\right) X , \\ R_{V+Y+sA} J_X V &= -\frac{1}{4}(1 + 3|V|^2) J_X V - \frac{3}{4}|V|^2 |Y| K X - \frac{3}{4} s |V|^2 X .\end{aligned}$$

So we see that $\mathfrak{q}_j$ is invariant under R_{V+Y+sA}.

We continue with the case $\mu_k = -1$ and put

$$\tilde{V} := J_Y V - sV .$$

Then we get for each non-zero $X \in L_k$

$$\begin{aligned} R_{V+Y+sA}X &= \left(\frac{3}{4}|V|^2 - 1\right) X + \frac{3}{4} J_X \tilde{V} , \\ R_{V+Y+sA} J_X \tilde{V} &= \frac{3}{4}|V|^2(s^2 + |Y|^2)X - \frac{1}{4}(1 + 3|V|^2) J_X \tilde{V} . \end{aligned}$$

Thus $\text{span}\{X, J_X\tilde{V}\}$ is invariant under R_{V+Y+sA}. Hence there exist non-zero $\alpha, \beta \in \mathbb{R}$ and $\kappa \in \mathbb{R}$ such that

$$R_{V+Y+sA}(\alpha X + \beta J_X \tilde{V}) = \kappa\alpha X + \kappa\beta J_X \tilde{V} .$$

Comparing coefficients leads to the equations

$$\begin{aligned} \kappa\alpha &= \left(\frac{3}{4}|V|^2 - 1\right)\alpha + \frac{3}{4}|V|^2(s^2 + |Y|^2)\beta , \\ \kappa\beta &= \frac{3}{4}\alpha - \frac{1}{4}(1 + 3|V|^2)\beta . \end{aligned}$$

We may normalize β by putting $\beta := 3/4$. Then the second equation yields

$$\alpha = \kappa + \frac{1}{4} + \frac{3}{4}|V|^2 .$$

Inserting α and β into the first equation then gives

$$(\kappa + 1)\left(\kappa + \frac{1}{4}\right) = 0 .$$

Thus $\kappa \in \{-1, -1/4\}$ is an eigenvalue of $R_{V+Y+sA}|\mathfrak{q}_k$. The corresponding eigenspace is

$$\begin{aligned} \{|V|^2 X + J_X\tilde{V} \mid X \in L_k\} \quad &, \quad \text{if } \kappa = -1/4 , \\ \{(|V|^2 - 1)X + J_X\tilde{V} \mid X \in L_k\} \quad &, \quad \text{if } \kappa = -1 . \end{aligned}$$

Next, we consider the case $\mu_j \neq -1$.

We start with the subcase $j = 0$, $\mu_0 = 0$ and $s = 0$. For $X \in L_0$ we have

$$R_{V+Y} J_X V = -\frac{1}{4}(1 + 3|V|^2) J_X V .$$

Thus $-(1 + 3|V|^2)/4$ is an eigenvalue of R_{V+Y} and $J_{L_0}V$ is contained in the corresponding eigenspace. Further, we know that

$$\begin{aligned} R_{V+Y}X &= \frac{3}{4} J_X J_Y V - \left(1 - \frac{3}{4}|V|^2\right) X , \\ R_{V+Y} J_X J_Y V &= -\frac{1}{4} J_X J_Y V + \frac{3}{4}|V|^2|Y|^2 X . \end{aligned}$$

Thus, span$\{X, J_X J_Y V\}$ is invariant under R_{V+Y}. Hence there exist some non-zero $\alpha, \beta \in \mathbb{R}$ so that $\alpha X + \beta J_X J_Y V$ is an eigenvector of R_{V+Y}, say

$$R_{V+Y}(\alpha X + \beta J_X J_Y V) = \kappa\alpha X + \kappa\beta J_X J_Y V \ .$$

Then we get the equations

$$\begin{aligned} \kappa\alpha &= \left(\frac{3}{4}|V|^2 - 1\right)\alpha + \frac{3}{4}|V|^2|Y|^2\beta \ , \\ \kappa\beta &= \frac{3}{4}\alpha - \frac{1}{4}\beta \ . \end{aligned}$$

The second equation then gives

$$\left(\kappa + \frac{1}{4}\right)\beta = \frac{3}{4}\alpha \ ,$$

and normalizing β by $\beta := 3/4$, we deduce

$$\alpha = \kappa + \frac{1}{4} \ .$$

Thus the first equation implies

$$\left(\kappa + \frac{1}{4}\right)\left(\kappa + 1 - \frac{3}{4}|V|^2\right) = \frac{9}{16}|V|^2|Y|^2 \ .$$

The solutions of this quadratic equation are

$$\frac{1}{8}(3|V|^2 - 5 \pm 3|Y|\sqrt{1 + 3|V|^2}) \ ,$$

and the corresponding eigenspaces are given by

$$\{(4\kappa + 1)X + 3J_X J_Y V \mid X \in L_0\} \ .$$

It is easy to check that all three eigenvalues satisfy the equation

$$(\kappa + 1)\left(\kappa + \frac{1}{4}\right)^2 = \frac{27}{64}|V|^4|Y|^2 \ .$$

Moreover, two of these eigenvalues coincide, namely $-(1+3|V|^2)/4$ and $(3|V|^2 - 5 - 3|Y|\sqrt{1+3|V|^2})/8$, precisely if $|V|^2 = 2/3$.

We now treat the remaining cases. Let X be a non-zero vector in L_j and put

$$\tilde{V} := J_Y V - sV \ .$$

Then we obtain

$$\begin{aligned} R_{V+Y+sA} X &= \left(\frac{3}{4}|V|^2 - 1\right)X + \frac{3}{4}J_X\tilde{V} \ , \\ R_{V+Y+sA} J_X\tilde{V} &= \frac{3}{4}|V|^2(s^2 + |Y|^2)X - \frac{1}{4}J_X\tilde{V} - \frac{3}{4}|V|^2 J_{|Y|KX-sX}V \ , \\ R_{V+Y+sA} J_{|Y|KX-sX}V &= \frac{3}{4}|V|^2(s^2 - \mu_j|Y|^2)X - \frac{1}{4}(1+3|V|^2)J_{|Y|KX-sX}V \ . \end{aligned}$$

Now in the case we are considering, the vectors X, $J_X\tilde{V}$ and $J_{|Y|KX-sX}V$ are linearly independent. We also see that the span of these three vectors is invariant under R_{V+Y+sA}. In order to compute the eigenvalues and eigenvectors we define a non-zero vector E by

$$E := \alpha X + \beta J_X\tilde{V} + \delta J_{|Y|KX-sX}V\ .$$

Then the eigenvector equation

$$R_{V+Y+sA}E = \kappa E$$

yields the following system of equations:

$$\begin{aligned}
\kappa\alpha &= \left(\frac{3}{4}|V|^2 - 1\right)\alpha + \frac{3}{4}|V|^2(s^2 + |Y|^2)\beta + \frac{3}{4}|V|^2(s^2 - \mu_j|Y|^2)\delta\ ,\\
\kappa\beta &= \frac{3}{4}\alpha - \frac{1}{4}\beta\ ,\\
\kappa\delta &= -\frac{3}{4}|V|^2\beta - \frac{1}{4}(1 + 3|V|^2)\delta\ .
\end{aligned}$$

The second equation yields

$$\left(\kappa + \frac{1}{4}\right)\beta = \frac{3}{4}\alpha\ .$$

As $\beta \neq 0$ (otherwise also $\alpha = 0$ and thus, by the first equation, $s = 0 = |Y|$ or $\delta = 0$, which is impossible in this case), we may normalize β by $\beta := 3/4$ and get

$$\alpha = \kappa + \frac{1}{4}\ .$$

The third equation implies

$$\left(\kappa + \frac{1}{4} + \frac{3}{4}|V|^2\right)\delta = -\frac{3}{4}|V|^2\beta = -\frac{9}{16}|V|^2\ ,$$

which yields

$$\delta \neq 0 \neq \kappa + \frac{1}{4} + \frac{3}{4}|V|^2\ .$$

Eventually, we multiply the first equation with $\kappa + (1 + 3|V|^2)/4$ and obtain

$$\begin{aligned}
0 &= \kappa^3 + \frac{3}{2}\kappa^2 + \frac{9}{16}\kappa + \frac{1}{16} - \frac{27}{64}|V|^4|Y|^2(1 + \mu_j)\\
&= (\kappa + 1)\left(\kappa + \frac{1}{4}\right)^2 - \frac{27}{64}|V|^4|Y|^2(1 + \mu_j)\ .
\end{aligned}$$

Thus each eigenvalue κ of $R_{V+Y+sA}|\mathfrak{q}_j$ is a solution of

$$(\kappa + 1)\left(\kappa + \frac{1}{4}\right)^2 = \frac{27}{64}|V|^4|Y|^2(1 + \mu_j)\ ,$$

and the corresponding eigenspace is given by

$$\{(4\kappa{+}1)(4\kappa{+}1{+}3|V|^2)X + 3(4\kappa{+}1{+}3|V|^2)J_X\tilde{V} - 9|V|^2 J_{|Y|KX-sX}V \mid X \in L_j\}\ .$$

It remains to investigate whether some of these eigenvalues can be equal. Therefore we define the functions

$$f : \mathbb{R} \to \mathbb{R}\ ,\ x \mapsto (x+1)\left(x+\frac{1}{4}\right)^2\ ,$$
$$g : G \to \mathbb{R}\ ,\ (x,y) \mapsto \frac{27}{64}(1+\mu_j)x^4y^2\ ,$$
$$h : \mathbb{R} \to \mathbb{R}\ ,\ x \mapsto \frac{27}{64}(1+\mu_j)x^4(1-x^2)\ ,$$

where

$$G := \{(x,y) \in \mathbb{R}^2 \mid x, y \geq 0\ ,\ x^2+y^2 \leq 1\}\ .$$

The eigenvalues κ of $R_{V+Y+sA}|\mathfrak{q}_j$ are the solutions of

$$f(\kappa) = g(|V|,|Y|)\ .$$

The function f has two zeroes, namely at -1 and $-1/4$, the second one is of multiplicity two. The only local maximum of f is at $-3/4$ with value $1/16$, and the only local mimimum is at $-1/4$ with value 0. Clearly, g has no critical points in the interior of G and $g(x,y) = 0$ if $x = 0$ or $y = 0$. Hence the maximum of g must lie in the set

$$\{(x,y) \in \mathbb{R}^2 \mid x,y > 0\ ,\ x^2+y^2 = 1\}\ .$$

In order to find it we differentiate h and obtain

$$h'(x) = \frac{27}{32}(1+\mu_j)x^3(2-3x^2)\ .$$

Thus the critical points of h are at 0 and $\pm\sqrt{2/3}$. We have $h(0) = 0$ and $h(\sqrt{2/3}) = (1+\mu_j)/16 \leq 1/16$, and equality if and only if $\mu_j = 0$. So the maximal value of g is $(1+\mu_j)/16$ and is attained for $(x,y) = (\sqrt{2/3}, \sqrt{1/3})$. Combining this with the above discussion of the graph of f we conclude that all eigenvalues $\kappa_1, \kappa_2, \kappa_3$ are distinct unless $\mu_j = 0$, $s = 0$ and $|V|^2 = 2/3$. □

The Theorem shows that all eigenvalues of the Jacobi operator lie between -1 and 0. As the minimal and maximal value of the sectional curvature $K(\sigma)$ of S arise as eigenvalues of the Jacobi operator we conclude that $-1 \leq K(\sigma) \leq 0$. Furthermore, -1 always arises as an eigenvalue, whence -1 is the minimal value of the sectional curvature of S. It can also be seen immediately that the maximal value of $K(\sigma)$ is $-1/4$ provided that $\mathfrak{n}$ satisfies the J^2-condition, that is (see 4.1.9), if S is a symmetric space. If $\mathfrak{n}$ does not satisfy the J^2-condition, the maximal value of the sectional curvature of S is not known to us in general. J. Boggino [Bog] stated that every non-symmetric Damek-Ricci space admits a two-plane σ for which $K(\sigma)$ vanishes. But a careful check shows that the sectional curvature of S with respect to the plane σ written down by Boggino in the proof of Theorem 2 does not vanish except in the situation described in the following. The preceding Theorem shows that zero occurs as a value of $K(\sigma)$ if and only if there exists a unit vector $V+Y \in \mathfrak{n}$ with $|V|^2 = 2/3$ such that zero is an eigenvalue of the operator $K^2_{V,Y}$. Recall that zero is an eigenvalue of $K^2_{V,Y}$ means that there exists a non-zero vector $X \in Y^\perp$ so that $J_X J_Y V$ is orthogonal to $J_{\mathfrak{z}} V$. Let $V + Y \in \mathfrak{n}$ be a unit vector with $V \neq 0 \neq Y$. Then we may decompose $\mathfrak{z}$ orthogonally into

$$\mathfrak{z} = \mathbb{R}Y \oplus L_0 \oplus L_1 \oplus \ldots \oplus L_k \ .$$

According to the Theorem, the dimension of L_j is even if $\mu_j \neq 0$. So if m, the dimension of $\mathfrak{z}$, is even, L_0 must be odd-dimensional and hence zero arises as an eigenvalue of $K^2_{V,Y}$. As $\mathfrak{a}$ is one-dimensional and $\mathfrak{v}$ is even-dimensional, it follows that m is even precisely if S is odd-dimensional. Summing up, we conclude

Proposition *The sectional curvatures $K(\sigma)$ of a Damek-Ricci space S satisfy $-1 \leq K(\sigma) \leq 0$ and -1 is attained. If S is symmetric then $-1 \leq K(\sigma) \leq -1/4$ and $-1/4$ is attained. S attains zero as a value of the sectional curvature if and only if there exist a unit vector $V + Y \in \mathfrak{n}$ with $|V|^2 = 2/3$ and a non-zero vector $X \in Y^{\perp}$ so that $J_X J_Y V$ is orthogonal to $J_{\mathfrak{z}} V$. If S is odd-dimensional, then zero is attained as a value of $K(\sigma)$.*

For the even-dimensional non-symmetric Damek-Ricci spaces we do not know whether zero arises as a value for the sectional curvature except in some special cases treated by E. Damek [Dam1], where also another detailed account on the sectional curvature of these spaces can be found.

4.3 Eigenvalues and eigenvectors along geodesics

In this section we will prove that the eigenvalues of the Jacobi operator of a Damek-Ricci space S are constant along geodesics precisely if S is symmetric. We will also show that S is symmetric if and only if the eigenspaces of the Jacobi operator along geodesics are invariant under parallel translation.

Theorem 1 *A Damek-Ricci space is a $\mathfrak{C}$-space if and only if it is a symmetric space.*

We will provide two alternative proofs of Theorem 1, one using the eigenvalues and the other one using eigenvectors of the Jacobi operator.

First Proof. If S is a symmetric Damek-Ricci space, then it is obviously a $\mathfrak{C}$-space. So, let S be a non-symmetric Damek-Ricci space. Then the corresponding generalized Heisenberg algebra $\mathfrak{n}$ does not satisfy the J^2-condition, which means that there exists a unit vector $V + Y \in \mathfrak{n}$ with $V \neq 0 \neq Y$ such that not all eigenvalues of $K^2_{V,Y}$ are equal to -1. Let $\gamma : \mathbb{R} \to S$ be the geodesic in S with $\gamma(0) = e$ and $\dot{\gamma}(0) = V + Y$. According to Theorem 2 in 4.1.11 we have

$$\dot{\gamma}(t) = V(t) + Y(t) + A(t)$$

with

$$V(t) = \frac{\sqrt{h(t)}}{\chi(t)}((1 - |Y|^2\theta^2)(t)V + 2\theta(t)J_Y V) \ ,$$

$$
\begin{aligned}
Y(t) &= h(t)Y \ , \\
A(t) &= (\ln h)'(t)A \ , \\
\theta(t) &= \tanh(t/2) \ , \\
\chi(t) &= 1+|Y|^2\theta^2(t) \ , \\
h(t) &= \frac{1-\theta^2(t)}{\chi(t)} \ .
\end{aligned}
$$

Clearly, $Y(t)^\perp = Y^\perp$ for all $t \in \mathbb{R}$. Thus the operators $K_{V(t),Y(t)}$ are well-defined on $Y^\perp$ for all $t \in \mathbb{R}$. A straightforward computation shows that

$$
[V(t), J_X J_Y V(t)] = h(t)[V, J_X J_Y V] \ ,
$$

and therefore, using $Y(t) = h(t)Y$, $|V(t)|^2 = |V|^2 h(t)$ and $|Y(t)| = |Y|h(t)$,

$$
K_{V(t),Y(t)}X = \frac{1}{|V(t)|^2|Y(t)|}[V(t), J_X J_{Y(t)}V(t)] = K_{V,Y}X \ .
$$

From this we conclude that the eigenvalues of $K^2_{V(t),Y(t)}$ do not depend on t. Now, let $\mu \neq -1$ be an eigenvalue of $K^2_{V,Y}$. According to Theorem 4.2 there are some eigenvalues $\kappa(t)$ of the Jacobi operator $R_\gamma(t) = R_{V(t)+Y(t)+A(t)}$ satisfying the relation

$$
(\kappa(t)+1)\left(\kappa(t)+\frac{1}{4}\right)^2 = \frac{27}{64}|V(t)|^4|Y(t)|^2(1+\mu) = \frac{27}{64}|V|^4|Y|^2h^4(t)(1+\mu) \ .
$$

As $1+\mu \neq 0$ and h is non-constant, the zeroes of this third order equation cannot be constant. This shows that R_γ has non-constant eigenvalues and hence S is not a $\mathfrak{C}$-space. □

Second Proof. If S is a symmetric Damek-Ricci space, then it is a $\mathfrak{C}$-space. From now on we assume that S is a non-symmetric Damek-Ricci space. Then the corresponding generalized Heisenberg algebra $\mathfrak{n}$ does not satisfy the J^2-condition. This implies that there exists a unit vector $V+Y+sA \in \mathfrak{s}$ with $V \neq 0 \neq Y$ and $s \neq 0$ such that not all eigenvalues of $K := K^2_{V,Y}$ are equal to -1. Let μ be such an eigenvalue. According to Theorem 4.2 there exists an eigenvalue κ of R_{V+Y+sA} such that

$$
R_{V+Y+sA}E = \kappa E
$$

with

$$
E := (4\kappa+1)(4\kappa+1+3|V|^2)X + 3(4\kappa+1+3|V|^2)J_X(J_Y V - sV) - 9|V|^2 J_{|Y|KX-sX}V \ ,
$$

where X is some unit eigenvector of K^2 with respect to μ. Using the formula for R'_{V+Y+sA} in 4.1.8 we obtain

$$
\begin{aligned}
\frac{2}{3}R'_{V+Y+sA}E &= -3|V|^2|Y|^2(4\kappa+1+3|V|^2(1+\mu))J_X V \\
&\quad -3|V|^2|Y|s(4\kappa+1)J_{KX}V \\
&\quad +3|V|^2 s(4\kappa+1)J_X J_Y V \\
&\quad -3|V|^2|Y|(4\kappa+1)J_{KX}J_Y V \ .
\end{aligned}
$$

We write E also in this form, that is,

$$\begin{aligned} E &= -3s(4\kappa+1)J_X V \\ &\quad -9|V|^2|Y|J_{KX}V \\ &\quad +3(4\kappa+1+3|V|^2)J_X J_Y V \\ &\quad +(4\kappa+1)(4\kappa+1+3|V|^2)X . \end{aligned}$$

The vectors

$$J_X V\ ,\ J_{KX}V\ ,\ J_X J_Y V\ ,\ J_{KX}J_Y V\ ,\ X$$

are pairwise orthogonal to each other except in the cases when

$$\begin{aligned} <J_{KX}V, J_X J_Y V> &= -|V|^2|Y|\mu , \\ <J_X V, J_{KX}J_Y V> &= |V|^2|Y|\mu . \end{aligned}$$

Considering this, a straightforward computation then yields

$$\frac{2}{3}<R'_{V+Y+sA}E, E> = 18|V|^4|Y|^2 s(1+\mu)(4\kappa+1)(4\kappa+1+3|V|^2) .$$

By the assumption $|V|$, $|Y|$, s, and $1+\mu$ are non-zero. As we are in the case (vi)(B) of Theorem 4.2, κ cannot be equal to $-1/4$, whence $4\kappa+1 \neq 0$. If $4\kappa+1+3|V|^2 = 0$, then $J_{|Y|KX-sX}V$ would be an eigenvector of R_{V+Y+sA}. But we have

$$R_{V+Y+sA}J_{|Y|KX-sX}V = \frac{3}{4}|V|^2(s^2-\mu|Y|^2)X - \frac{1}{4}(1+3|V|^2)J_{|Y|KX-sX}V ,$$

and since $\mu \leq 0$, the X-component does not vanish. So we conclude that

$$<R'_{V+Y+sA}E, E> \neq 0 ,$$

which shows that condition (iii) in Proposition 2 of 2.8 does not hold. Hence S cannot be a $\mathfrak{C}$-space. □

The second main result of this section is

Theorem 2 *A Damek-Ricci space S is a $\mathfrak{P}$-space if and only if it is a symmetric space.*

Proof. If S is symmetric, then it is also a $\mathfrak{P}$-space. Suppose now that S is a $\mathfrak{P}$-space. Let $V+Y+sA \in \mathfrak{s}$ be a unit vector with $V \neq 0 \neq Y$ and $s \neq 0$, $K := K_{V,Y}$, and $0 \neq X \in Y^\perp$. According to Proposition 3 in 2.8 we have

$$[R_{V+Y+sA}, R'_{V+Y+sA}] = 0 .$$

Since

$$R'_{V+Y+sA}X = 0 ,$$

$$\begin{aligned} R'_{V+Y+sA}R_{V+Y+sA}X &= R'_{V+Y+sA}\left(\frac{3}{4}J_X J_Y V - \frac{3}{4}sJ_X V + \left(\frac{3}{4}|V|^2-1\right)X\right) \\ &= \frac{9}{8}|V|^2(s(J_X J_Y V - J_{|Y|KX}V) - |Y|(J_{KX}J_Y V + |Y|J_X V)) , \end{aligned}$$

and

$$J_X J_Y V - J_{|Y|KX}V \ , \ J_{KX}J_YV + |Y|J_XV$$

are orthogonal to each other, we conclude that

$$J_X J_Y V = J_{|Y|KX}V \ \text{ for all } X \in Y^\perp \ .$$

Thus the corresponding generalized Heisenberg algebra $\mathfrak{n}$ satisfies the J^2-condition and hence S is a symmetric space. □

4.4 Harmonicity

In this section we provide an alternative proof of the result by E. Damek and F. Ricci [DaRi1] stating that S is always a harmonic space. We start with a Lemma providing a formula for the Laplacian Δ = div grad of S in terms of the global coordinates on S introduced in 4.1.5.

Lemma *The Laplace operator Δ of S is given by*

$$\begin{aligned} \Delta &= e^\lambda \sum_i \frac{\partial^2}{\partial v_i^2} + e^\lambda \left(e^\lambda + \frac{1}{4}\sum_i v_i^2\right) \sum_i \frac{\partial^2}{\partial y_i^2} + \frac{\partial^2}{\partial \lambda^2} \\ &\quad - \left(m + \frac{n}{2}\right)\frac{\partial}{\partial \lambda} + \frac{1}{2}e^\lambda \sum_{i,j,k} A_{ij}^k v_i \frac{\partial}{\partial v_j}\frac{\partial}{\partial y_k} \ . \end{aligned}$$

Proof. Using the notations in 4.1.5 we have

$$\begin{aligned} \Delta &= \sum_i V_i^2 + \sum_i Y_i^2 + A^2 - \sum_i \nabla_{V_i} V_i - \sum_i \nabla_{Y_i} Y_i - \nabla_A A \\ &= \sum_i V_i^2 + \sum_i Y_i^2 + A^2 - \left(m + \frac{n}{2}\right) A \ . \end{aligned}$$

Inserting here the expressions for V_i, Y_i and A according to Lemma 4.1.5 yields the assertion by a lengthy but straightforward computation. □

The next step is to derive a formula for the differentiable function

$$\Omega_e := \frac{1}{2} d_e^2$$

on S, where $d_e(p)$ is the distance from $p \in S$ to e, in terms of these coordinates.

Proposition 1 *We have*

$$\Omega_e = \Phi \circ \Psi \ ,$$

where Φ is the diffeomorphism

$$\Phi : [4, \infty[\to [0, \infty[\ , \ t \mapsto 2\,\mathrm{Artanh}^2\left(\sqrt{1 - \frac{4}{t}}\right)$$

and

$$\Psi := e^{-\lambda}\left(\left(1+\frac{1}{4}\sum_i v_i^2+e^{\lambda}\right)^2+\sum_i y_i^2\right) : S \to [4,\infty[\ .$$

Proof. Let $p := (\exp_{\mathfrak{n}}(U+X), r) \in S$ be arbitrary, but not equal to e. As S is complete, simply connected, and of non-positive curvature, there is a unique geodesic $\gamma : \mathbb{R} \to S$ in S which is parametrized by arc length and satisfies $\gamma(0) = e$ and $\gamma(t) = p$ for some $t \in \mathbb{R}_+$. Clearly, we have $d_e(p) = t$. If $V + Y + sA := \dot\gamma(0)$, then, by means of Theorem 1 in 4.1.11, we obtain the relations

$$\begin{aligned} U &= \frac{2\theta(1-s\theta)}{\chi}(t)V + \frac{2\theta^2}{\chi}(t)J_Y V\ , \\ X &= \frac{2\theta}{\chi}(t)Y\ , \\ r &= \ln\left(\frac{1-\theta^2}{\chi}\right)(t)\ , \end{aligned}$$

with

$$\theta(t) = \tanh(t/2)\ ,\ \chi(t) = (1-s\theta)^2(t) + |Y|^2\theta^2(t)\ .$$

This implies

$$|U|^2 = \frac{4\theta^2}{\chi}(t)|V|^2\ ,\ |X|^2 = \frac{4\theta^2}{\chi^2}(t)|Y|^2\ ,\ \theta^2(t) = 1-\chi(t)e^r\ ,$$

and therefore,

$$\chi(t) = (1-s\theta)^2(t) + |Y|^2\theta^2(t) = (1-s\theta)^2(t) + \frac{\chi^2}{4}|X|^2$$

and

$$1 = |V|^2 + |Y|^2 + s^2 = \frac{\chi}{4\theta^2}(t)|U|^2 + \frac{\chi^2}{4\theta^2}(t)|X|^2 + s^2\ .$$

Multiplication of the last equation with $\theta^2(t)$ and comparing it with the preceding one yields

$$\chi(t)\left(1+\frac{1}{4}|U|^2\right) = 1 - 2s\theta(t) + \theta^2(t)\ .$$

We now replace $\theta^2(t)$ by $1-\chi(t)e^r$ and obtain

$$\chi(t)\left(1+\frac{1}{4}|U|^2+e^r\right) = 2(1-s\theta)(t)\ .$$

Squaring this equation, then replacing $(1-s\theta)^2(t)$ by $(\chi - \chi^2|X|^2/4)(t)$, and then dividing by $\chi(t)$ yields

$$\chi(t)\left(\left(1+\frac{1}{4}|U|^2+e^r\right)^2+|X|^2\right) = 4\ .$$

Inserting the resulting expression for $\chi(t)$ in

$$d_e(p) = t = 2\mathrm{Artanh}(\theta(t)) = 2\mathrm{Artanh}\left(\sqrt{1-\chi(t)e^r}\right)$$

then yields the required formula for Ω_e. The fact that Φ is a diffeomorphism from $[4,\infty[$ onto $[0,\infty[$, and the statement that Ψ maps S onto $[4,\infty[$, can be proved easily. □

Before we come to harmonicity we recall the notion of an isoparamatric function. A differentiable function $f : M \to \mathbb{R}$ on a Riemannian manifold M is called *isoparametric* if

$$\|\mathrm{grad}\, f\|^2 = \alpha \circ f \quad \text{and} \quad \Delta f = \beta \circ f$$

with some continuous functions α, β. The theory of isoparametric hypersurfaces (see 4.1.10) originates from a result by E. Cartan [Car] stating that a function on a space of constant curvature is isoparametric if and only if its level hypersurfaces have constant principal curvatures. Cartan's result is not true in more general spaces as was shown by Q.M. Wang [Wan1] by providing an isoparametric function on complex projective space whose level hypersurfaces have non-constant principal curvatures. The geometric meaning of $\|\mathrm{grad}\, f\|^2 = \alpha \circ f$ is that locally the level hypersurfaces of f are equidistant (see for instance [Wan2] for a proof). Functions satisfying only this condition are also called *transnormal*. For a transnormal function f the equation $\Delta f = \beta \circ f$ is equivalent to the geometrical property that the level hypersurfaces of f have constant mean curvature (see for instance [Nom]).

Proposition 2 *The function Ω_e is an isoparametric function on S.*

Proof. As

$$\mathrm{grad}\,(\Phi \circ \Psi) = (\Phi' \circ \Psi)\mathrm{grad}\,\Psi$$

and

$$\Delta(\Phi \circ \Psi) = (\Phi'' \circ \Psi)\|\mathrm{grad}\,\Psi\|^2 + (\Phi' \circ \Psi)\Delta\Psi \ ,$$

it suffices to prove that Ψ is an isoparametric function; and this can be checked without difficulty by using the expression for Δ in the preceding Lemma. Explicitly, it turns out that

$$\begin{aligned} \|\mathrm{grad}\ \Psi\|^2 &= \Psi^2 - 4\Psi \ , \\ \Delta\Psi &= \left(1 + \frac{n}{2} + m\right)\Psi - 2(m+1) \ . \end{aligned}$$

So the assertion is proved. □

The level hypersurfaces of Ω_e are the geodesic spheres in S centered at e. The preceding Proposition 2, together with the homogeneity of S, therefore implies that the geodesic spheres in S have constant mean curvature. We now apply characterization (iv) for harmonic spaces in Proposition 1 of 2.6 and eventually obtain

Theorem [DaRi1] *Every Damek-Ricci space is a harmonic space.*

By a look at the table for the dimensions of generalized Heisenberg groups in 3.1.2 it follows that the lowest dimension of a non-symmetric Damek-Ricci space is seven. It is still an open problem whether there exist non-symmetric harmonic spaces in dimensions

$$5\ ,\ 6\ ,\ 8\ ,\ 9\ ,\ 10\ ,\ 17\ ,\ 18\ ,\ 26\ ,\ldots$$

and all other higher dimensions in which there are no non-symmetric Damek-Ricci spaces.

4.5 Geometrical consequences

We will now draw some conclusions from the results obtained in 4.3 and 4.4. According to B.Y. Chen and the third author [ChVa, Theorem 6.22] every non-flat harmonic space is irreducible as a Riemannian manifold. So a consequence of the harmonicity of the Damek-Ricci spaces, and the fact that these spaces are non-flat, is

Theorem 1 *Every Damek-Ricci space is irreducible as a Riemannian manifold.*

Several of the classes of Riemannian manifolds introduced in Section 2 form a subclass of the $\mathfrak{C}$-spaces. So, as a consequence of Theorem 1 in 4.3 we obtain

Theorem 2 *For a Damek-Ricci space S the following statements are equivalent:*

(i) S is a symmetric space;

(ii) S is a naturally reductive Riemannian homogeneous space;

(iii) S is a Riemannian g.o. space;

(iv) S is a weakly symmetric space;

(v) S is a commutative space;

(vi) S is a $\mathfrak{C}_0$-space;

(vii) S is a $\mathfrak{TC}$-space;

(viii) S is an $\mathfrak{SC}$-space;

(ix) S is a globally Osserman space.

Using the irreducibilty of Damek-Ricci spaces the equivalence of (i), (ii), (iii) and (v) can also be obtained from more general results about homogeneous Hadamard manifolds. For (iii), and hence also (ii), we refer to [Wol2], and for (v) see the remark at the end of [Wol1].

Another consequence of the harmonicity of Damek-Ricci spaces is

Theorem 3 *Every Damek-Ricci space is*

(i) a probabilistic commutative space;

(ii) a D'Atri space.

So, all the Damek-Ricci spaces belonging to the various classes introduced in Section 2 have now been completely determined. As consequences of the preceding two theorems we obtain

Corollary 1 *Every non-symmetric Damek-Ricci space is a D'Atri space which is not a $\mathfrak{C}$-space.*

Corollary 2 *Every non-symmetric Damek-Ricci space is a probabilistic commutative space which is not commutative.*

Two our knowledge these two corollaries provide the first examples showing that the classes of $\mathfrak{C}$- and D'Atri spaces, and the classes of commutative and probabilistic commutative spaces, respectively, do not coincide. It is still an open problem whether any $\mathfrak{C}$-space is a D'Atri space or not.

Of great interest is also the question whether there exist compact quotients of Damek-Ricci spaces. In this regard a general result of R. Azencott and E. Wilson [AzWi] implies

Theorem 4 *A Damek-Ricci space admits a quotient of finite volume if and only if it is a symmetric space.*

We recall that a Riemannian manifold M is called *semi-symmetric* if its curvature tensor R satisfies $R(X,Y) \cdot R = 0$ for all vector fields X, Y on M, where $R(X,Y)$ acts on R as a derivation. Clearly, every symmetric space is also semi-symmetric. According to E. Boeckx [Boe1], every semi-symmetric Riemannian manifold whose Ricci tensor is a Killing tensor is locally symmetric. Now, every Damek-Ricci space is an Einstein manifold and hence its Ricci tensor is obviously a Killing tensor. Thus we derive

Theorem 5 *A Damek-Ricci space is semi-symmetric if and only if it is symmetric.*

A nice geometrical problem, whose treatment has been developed by B.Y. Chen and the third author in [ChVa], is to study the geometry of Riemannian manifolds in terms of properties of their small geodesic spheres. Let us recall that a Riemannian manifold M is called *curvature-homogeneous* if at any two points $p, q \in M$ the Riemannian curvature tensors coincide via a linear isometry from T_pM onto T_qM. In [BePrVa2] it was proved that a Riemannian manifold, whose small geodesic spheres are curvature-homogeneous, is a harmonic globally Osserman space. As geodesic spheres in the symmetric Damek-Ricci spaces are even homogeneous, this and Theorem 2 imply

Theorem 6 *All geodesic spheres in a Damek-Ricci space S are curvature-homogeneous if and only if S is symmetric.*

The harmonicity of S just means that the geodesic spheres in S have constant mean curvature. Assume that the geodesic spheres in S are isoparametric. Then, obviously, the principal curvatures of geodesic spheres in S are the same at antipodal points, whence S is an $\mathfrak{SC}$-space. Furthermore, the two-point homogeneity of the symmetric Damek-Ricci spaces implies immediately that their geodesic spheres are isoparametric. Thus we conclude

Theorem 7 [TrVa3] *All geodesic spheres in a Damek-Ricci space S are isoparametric if and only if S is symmetric.*

So the distance function Ω_e (see 4.4) on a non-symmetric Damek-Ricci space provides another example of an isoparametric function whose level hypersurfaces do not have constant principal curvatures.

The long-standing conjecture about harmonic spaces, reformulated in terms of geodesic spheres, stated that a Riemannian manifold is locally isometric to a two-point homogeneous space if and only if all its small geodesic spheres have constant mean curvature. The non-symmetric Damek-Ricci spaces provide counterexamples to this conjecture, but the preceding theorem shows that the principal curvatures of small geodesic spheres in these spaces are non-constant. This motivated the second and third author to formulate the

Conjecture [TrVa3] *A Riemannian manifold is locally isometric to a two-point homogeneous space if and only if all its small geodesic spheres are isoparametric.*

As any space, whose small geodesic spheres are isoparametric, is a globally Osserman space [GiSwVa], the conjecture is known to be true in many dimensions and cases (see 2.12 for details). For a further discussion of this conjecture we refer to [TrVa3].

For a treatment of an intrinsic version of the above conjecture by using the Ricci tensor of the small geodesic spheres instead of their shape operator, see [GiSwVa].

Moreover, Theorem 4.4 also yields that the Damek-Ricci spaces provide counterexamples for other conjectures concerning harmonic spaces, more precisely, concerning k-harmonic spaces (see [TrVa3], [TrVa4], [Van2], [Wil]).

Although the Damek-Ricci examples, with their rich and interesting geometry, provide a negative answer to the fundamental conjecture about harmonic manifolds, they are not at the end of a long story. On the contrary, they give rise to many yet unsolved questions and one may hope and believe that the research about these problems will lead to beautiful geometrical results. We finish this section by mentioning just a few of them:

- Do there exist harmonic spaces which are not locally symmetric and not locally isometric to a Damek-Ricci space? Find explicit examples.

- Are harmonic spaces always locally homogeneous?

- Classify the harmonic Hadamard manifolds, homogeneous or not.

- Classify the harmonic spaces with strictly negative sectional curvature.

- Do there exist non-flat Ricci flat harmonic spaces?

- Classify the harmonic spaces and study their geometry in more detail.

Some research about these questions has already been started. For example, in [Heb] the author started the study of Hadamard manifolds which are D'Atri spaces and obtained several results relating to the theory about harmonic spaces. See also [BeCoGa] concerning the fourth problem in the compact case.

Bibliography

[Ale] D.V. Alekseevskii: Classification of quaternionic spaces with a transitive solvable group of motions. *Math. USSR-Izv.* **9** (1975), 297-339.

[AmSi] W. Ambrose, I.M. Singer: On homogeneous Riemannian manifolds. *Duke Math. J.* **25** (1958), 647-669.

[AtBoSh] M.F. Atiyah, R. Bott, A. Shapiro: Clifford modules. *Topology* **3**, Suppl. 1 (1964), 3-38.

[AzWi] H. Azencott, E. Wilson: *Homogeneous manifolds with negative curvature II.* Mem. Amer. Math. Soc. **8**, 178 (1976).

[BePrVa1] J. Berndt, F. Prüfer, L. Vanhecke: Symmetric-like Riemannian manifolds and geodesic symmetries. To appear in *Proc. Roy. Soc. Edinburgh Sect. A.*

[BePrVa2] J. Berndt, F. Prüfer, L. Vanhecke: Geodesic spheres and two-point homogeneous spaces. To appear in *Israel J. Math.*

[BeTrVa] J. Berndt, F. Tricerri, L. Vanhecke: Geometry of generalized Heisenberg groups and their Damek-Ricci harmonic extensions. *C.R. Acad. Sci. Paris Sér. I Math.* **318** (1994), 471-476.

[BeVa1] J. Berndt, L. Vanhecke: Two natural generalizations of locally symmetric spaces. *Diff. Geom. Appl.* **2** (1992), 57-80.

[BeVa2] J. Berndt, L. Vanhecke: Geodesic sprays and $\mathfrak{C}$- and $\mathfrak{P}$-spaces. *Rend. Sem. Mat. Univ. Politec. Torino* **50** (1992), 343-358.

[BeVa3] J. Berndt, L. Vanhecke: Geodesic spheres and generalizations of symmetric spaces. *Boll. Un. Mat. Ital. A (7)* **7** (1993), 125-134.

[BeVa4] J. Berndt, L. Vanhecke: Naturally reductive Riemannian homogeneous spaces and real hypersurfaces in complex and quaternionic space forms. *Differential Geometry and Its Applications*, Proc. Conf. Opava/Czechoslovakia 1992 (Eds. O. Kowalski and D. Krupka), Silesian University, Opava, 1993, 353-364.

[BeVa5] J. Berndt, L. Vanhecke: Geometry of weakly symmetric spaces. Preprint, 1994.

[Bes1] A.L. Besse: *Manifolds all of whose geodesics are closed.* Ergeb. Math. Grenzgeb. 93, Springer-Verlag, Berlin, Heidelberg, New York, 1978.

[Bes2] A.L. Besse: *Einstein manifolds.* Ergeb. Math. Grenzgeb. (3) 10, Springer-Verlag, Berlin, Heidelberg, New York, 1987.

[BeCoGa] G. Besson, G. Courtois, S. Gallot: Volumes, entropies et rigidités des espaces localement symétriques de courbure strictement négative. *C.R. Acad. Sci. Paris Sér. I Math.* **319** (1994), 81-84.

[Bie] L. Bieszk: On natural reductivity of five-dimensional commutative spaces. *Note Mat.* **8** (1988), 13-43.

[BlVa] D.E. Blair, L. Vanhecke: New characterization of φ-symmetric spaces. *Kodai Math. J.* **10** (1987), 102-107.

[Boe1] E. Boeckx: Einstein-like semi-symmetric spaces. *Arch. Math. (Brno)* **29** (1993), 235-240.

[Boe2] E. Boeckx: Semi-symmetric $\mathfrak{P}$-spaces. To appear in *Comment. Math. Univ. Carolinae.*

[Bog] J. Boggino: Generalized Heisenberg groups and solvmanifolds naturally associated. *Rend. Sem. Mat. Univ. Politec. Torino* **43** (1985), 529-547.

[Cao] J. Cao: Rigidity for non-compact surfaces of finite area and certain Kähler manifolds. To appear in *Ergodic Theory Dynamical Systems.*

[Car] E. Cartan: Familles de surfaces isoparamétriques dans les espaces à courbure constante. *Ann. Mat. Pura Appl. (4)* **17** (1938), 177-191.

[Cha] I. Chavel: Isotropic Jacobi fields, and Jacobi's equations on Riemannian homogeneous spaces. *Comment. Math. Helv.* **42** (1967), 237-248.

[ChVa] B.Y. Chen, L. Vanhecke: Differential geometry of geodesic spheres. *J. Reine Angew. Math.* **325** (1981), 28-67.

[Chi1] Q.S. Chi: A curvature characterization of certain locally rank-one symmetric spaces. *J. Differential Geom.* **28** (1988), 187-202.

[Chi2] Q.S. Chi: Quaternionic Kähler manifolds and a curvature characterization of two-point homogeneous spaces. *Illinois J. Math.* **35** (1991), 408-418.

[Chi3] Q.S. Chi: Curvature characterization and classification of rank-one symmetric spaces. *Pacific J. Math.* **150** (1991), 31-42.

[Cho] J.T. Cho: Natural generalizations of locally symmetric spaces. *Indian J. Pure Appl. Math.* **24** (1993), 231-240.

[ChSeVa] J.T. Cho, K. Sekigawa, L. Vanhecke: Volume-preserving geodesic symmetries on four-dimensional Hermitian Einstein spaces. Preprint, 1994.

[Cor] V. Cortés: Alekseevskian spaces. To appear in *Diff. Geom. Appl.*

[CoDoKoRi] M. Cowling, A.H. Dooley, A. Korányi, F. Ricci: H-type groups and Iwasawa decompositions. *Adv. in Math.* **87** (1991), 1-41.

[Dam1] E. Damek: Curvature of a semi-direct extension of a Heisenberg type nilpotent group. *Colloq. Math.* **53** (1987), 249-253.

[Dam2] E. Damek: The geometry of a semi-direct extension of a Heisenberg type nilpotent group. *Colloq. Math.* **53** (1987), 255-268.

[DaRi1] E. Damek, F. Ricci: A class of nonsymmetric harmonic Riemannian spaces. *Bull. Amer. Math. Soc. (N.S.)* **27** (1992), 139-142.

[DaRi2] E. Damek, F. Ricci: Harmonic analysis on solvable extensions of H-type groups. *J. Geom. Anal.* **2** (1992), 213-248.

[Dat] J.E. D'Atri: Geodesic spheres and symmetries in naturally reductive spaces. *Michigan Math. J.* **22** (1975), 71-76.

[DaDo] J.E. D'Atri, I. Dotti Miatello: A characterization of bounded symmetric domains by curvature. *Trans. Amer. Math. Soc.* **276** (1983), 531-540.

[DaNi1] J.E. D'Atri, H.K. Nickerson: Divergence-preserving geodesic symmetries. *J. Differential Geom.* **3** (1969), 467-476.

[DaNi2] J.E. D'Atri, H.K. Nickerson: Geodesic symmetries in spaces with special curvature tensors. *J. Differential Geom.* **9** (1974), 251-262.

[DaZi] J.E. D'Atri, W. Ziller: *Naturally reductive metrics and Einstein metrics on compact Lie groups.* Mem. Amer. Math. Soc. **18**, 215 (1979).

[dWVP] B. de Wit, A. Van Proeyen: Special geometry, cubic polynomials and homogeneous quaternionic spaces. *Comm. Math. Phys.* **149** (1992), 307-333.

[Dor] J. Dorfmeister: Homogeneous Kähler manifolds admitting a transitive solvable group of automorphisms. *Ann. Sci. École Norm. Sup. (4)* **18** (1985), 143-188.

[DoNa] J. Dorfmeister, K. Nakajima: The fundamental conjecture for homogeneous Kähler manifolds. *Acta Math.* **161** (1988), 23-70.

[Ebe1] P. Eberlein: *Structure of manifolds of nonpositive curvature.* Surveys in geometry, Tokyo, 1985, lecture notes.

[Ebe2] P. Eberlein: Geometry of 2-step nilpotent groups with a left invariant metric. *Ann. Sci. École Norm. Sup. (4)* **27** (1994), 611-660.

[Ebe3] P. Eberlein: Geometry of 2-step nilpotent groups with a left-invariant metric. II. *Trans. Amer. Math. Soc.* **343** (1994), 805-828.

[Gel] I.M. Gelfand: Spherical functions on symmetric spaces. *Dokl. Akad. Nauk SSSR* **70** (1950), 5-8. Translation: *Amer. Math. Soc. Transl. (Ser. 2)* **37** (1964), 39-44.

[GiSwVa] P.B. Gilkey, A. Swann, L. Vanhecke: Isoparametric geodesic spheres and a conjecture of Osserman concerning the Jacobi operator. To appear in *Quart. J. Math. Oxford Ser. (2)*.

[GoGoVa1] J.C. González-Dávila, M.C. González-Dávila, L. Vanhecke: Killing-transversally symmetric spaces. *Proc. Workshop Recent Topics in Differential Geometry Puerto de la Cruz* 1990 (Eds. D. Chinea and J.M. Sierra), Secret. Public. Univ. de La Laguna, Serie Informes 32, 1991, 77-88.

[GoGoVa2] J.C. González-Dávila, M.C. González-Dávila, L. Vanhecke: Reflections and isometric flows. To appear in *Kyungpook Math. J.*

[Gor1] C. Gordon: Naturally reductive homogeneous Riemannian manifolds. *Canad. J. Math.* **37** (1985), 467-487.

[Gor2] C. Gordon: Isospectral closed Riemannian manifolds which are not locally isometric. *J. Differential Geom.* **37** (1993), 639-649.

[GrWi] A. Gray, T.J. Willmore: Mean-value theorems for Riemannian manifolds. *Proc. Roy. Soc. Edinburgh Sect. A* **92** (1982), 343-364.

[Heb] J. Heber: Homogeneous spaces of nonpositive curvature and their geodesic flow. To appear in *Internat. J. Math. Math. Sci.*

[Hei] E. Heintze: On homogeneous manifolds of negative curvature. *Math. Ann.* **211** (1974), 23-34.

[Hel1] S. Helgason: *Differential geometry, Lie groups, and symmetric spaces.* Academic Press, New York, 1978.

[Hel2] S. Helgason: *Groups and geometric analysis.* Academic Press, New York, 1984.

[Jim1] J.A. Jiménez: Existence of Hermitian n-symmetric spaces and of non-commutative naturally reductive spaces. *Math. Z.* **196** (1987), 133-139. Addendum: *Math. Z.* **197** (1988), 455-456.

[Jim2] J.A. Jiménez: Non-commutative naturally reductive spaces of odd-dimension. Preprint.

[Jim3] J.A. Jiménez: Stiefel manifolds and non-commutative φ-symmetric spaces. Preprint.

[Kap1] A. Kaplan: Fundamental solutions for a class of hypoelliptic PDE generated by composition of quadratic forms. *Trans. Amer. Math. Soc.* **258** (1980), 147-153.

[Kap2] A. Kaplan: Riemannian nilmanifolds attached to Clifford modules. *Geom. Dedicata* **11** (1981), 127-136.

[Kap3] A. Kaplan: On the geometry of groups of Heisenberg type. *Bull. London Math. Soc.* **15** (1983), 35-42.

[KaRi] A. Kaplan, F. Ricci: Harmonic analysis on groups of Heisenberg type. *Harmonic analysis*, Proc. Conf. Cortona/Italy 1982 (Eds. G. Mauceri, F. Ricci and G. Weiss), Lecture Notes in Math. **992** (1983), 416-435.

[KoNo] S. Kobayashi, K. Nomizu: *Foundations of differential geometry II.* Interscience Publishers, New York, 1969.

[Kow1] O. Kowalski: *Generalized symmetric spaces.* Lecture Notes in Math. **805**, Springer-Verlag, Berlin, Heidelberg, New York, 1980.

[Kow2] O. Kowalski: Some curvature identities for commutative spaces. *Czech. Math. J.* **32** (1982), 389-396.

[Kow3] O. Kowalski: Spaces with volume-preserving symmetries and related classes of Riemannian manifolds. *Rend. Sem. Mat. Univ. Politec. Torino*, Fascicolo Speciale Settembre 1983, 131-158.

[KoPr] O. Kowalski, F. Prüfer: On probabilistic commutative spaces. *Monatsh. Math.* **107** (1989), 57-68.

[KoPrVa] O. Kowalski, F. Prüfer, L. Vanhecke: D'Atri spaces. Preprint, 1994.

[KoVa1] O. Kowalski, L. Vanhecke: Four-dimensional naturally reductive homogeneous spaces. *Rend. Sem. Mat. Univ. Politec. Torino*, Fascicolo Speciale Settembre 1983, 223-232.

[KoVa2] O. Kowalski, L. Vanhecke: Opérateurs différentiels invariants et symétries géodésiques préservant le volume. *C.R. Acad. Sci. Paris Sér. I Math.* **296** (1983), 1001-1003.

[KoVa3] O. Kowalski, L. Vanhecke: Classification of four-dimensional commutative spaces. *Quart. J. Math. Oxford Ser. (2)* **35** (1984), 281-291.

[KoVa4] O. Kowalski, L. Vanhecke: A generalization of a theorem on naturally reductive homogeneous spaces. *Proc. Amer. Math. Soc.* **91** (1984), 433-435.

[KoVa5] O. Kowalski, L. Vanhecke: Classification of five-dimensional naturally reductive spaces. *Math. Proc. Cambridge Philos. Soc.* **97** (1985), 445-463.

[KoVa6] O. Kowalski, L. Vanhecke: Two-point functions on Riemannian manifolds. *Ann. Global Anal. Geom.* **3** (1985), 95-119.

[KoVa7] O. Kowalski, L. Vanhecke: Riemannian manifolds with homogeneous geodesics. *Boll. Un. Mat. Ital. B (7)* **5** (1991), 189-246.

[Lic1] A. Lichnerowicz: Groupes kählериens. *C.R. Acad. Sci. Paris Sér. I Math.* **310** (1990), 671-676.

[Lic2] A. Lichnerowicz: Les groupes kählériens. *Symplectic geometry and mathematical physics* (Eds. P. Donato, C. Duval, J. Elhadad and G.M. Tuynman), a collection of articles in honor of J.-M. Souriau, Progress in Mathematics 99, Birkhäuser, Basel, 1991, 245-259.

[Nag] S. Nagai: Naturally reductive Riemannian homogeneous structure on a homogeneous real hypersurface in complex space forms. To appear in *Boll. Un. Mat. Ital. A.*

[Nom] K. Nomizu: Elie Cartan's work on isoparametric families of hypersurfaces. *Differential geometry,* Proc. Symp. Pure Math. Stanford/USA 1973 (Eds. S.S. Chern and R. Osserman), Proc. Symp. Pure Math. **27** (1975), 191-200.

[Oss] R. Osserman: Curvature in the eighties. *Amer. Math. Monthly* **97** (1990), 731-756.

[Pru] F. Prüfer: *Mittelwertmethoden in D'Atri-Räumen unter besonderer Berücksichtigung der Herleitung von Poissonschen Summationsformeln für Räume konstanter positiver Krümmung.* Dissertation B, Karl-Marx-Universität Leipzig, 1990.

[Rag] M.S. Raghunathan: *Discrete subgroups of Lie groups.* Ergeb. Math. Grenzgeb. **68**, Springer-Verlag, Berlin, New York, 1972.

[Ric] F. Ricci: Commutative algebras of invariant functions on groups of Heisenberg type. *J. London Math. Soc. (2)* **32** (1985), 265-271.

[Rie1] C. Riehm: The automorphism group of a composition of quadratic forms. *Trans. Amer. Math. Soc.* **269** (1982), 403-414.

[Rie2] C. Riehm: Explicit spin representations and Lie algebras of Heisenberg type. *J. London Math. Soc. (2)* **29** (1984), 49-62.

[RoUr] P.H. Roberts, H.D. Ursell: Random walk on a sphere and on a Riemannian manifold. *Philos. Trans. Roy. Soc. London Ser. A* **252** (1960), 317-356.

[RuWaWi] H.S. Ruse, A.G. Walker, T.J. Willmore: *Harmonic spaces.* Cremonese, Roma, 1961.

[SeVa1] K. Sekigawa, L. Vanhecke: Volume-preserving geodesic symmetries on four-dimensional 2-stein spaces. *Kodai Math. J.* **9** (1986), 215-224.

[SeVa2] K. Sekigawa, L. Vanhecke: Volume-preserving geodesic symmetries on four-dimensional Kähler manifolds. *Differential geometry,* Proc. 2nd Symp. Peñiscola/Spain 1985 (Eds. A.M. Naveira, A. Ferrández and F. Mascaró), Lecture Notes in Math. **1209** (1986), 275-291.

[SeVa3] K. Sekigawa, L. Vanhecke: Symplectic geodesic symmetries on Kähler manifolds. *Quart. J. Math. Oxford (2)* **37** (1986), 95-103.

[Sel] A. Selberg: Harmonic analysis and discontinuous groups in weakly symmetric Riemannian spaces with applications to Dirichlet series. *J. Indian Math. Soc. (N.S.)* **20** (1956), 47-87.

[Sza1] Z.I. Szabó: The Lichnerowicz conjecture on harmonic manifolds. *J. Differential Geom.* **31** (1990), 1-28.

[Sza2] Z.I. Szabó: Spectral theory for operator families on Riemannian manifolds. *Differential geometry*, Proc. Summer Research Institute on Differential Geometry, University of California, Los Angeles 1990 (Eds. R. Greene and S.T. Yau), Proc. Sympos. Pure Math. **54** Part 3 (1993), 615-665.

[Ton] P. Tondeur: *Foliations on Riemannian manifolds.* Springer-Verlag, New York, 1988.

[ToVa] P. Tondeur, L. Vanhecke: Transversally symmetric Riemannian foliations. *Tôhoku Math. J.* **42** (1990), 307-317.

[TrVa1] F. Tricerri, L. Vanhecke: *Homogeneous structures on Riemannian manifolds.* London Math. Soc. Lecture Note Ser. **83** (1983), Cambridge University Press, Cambridge.

[TrVa2] F. Tricerri, L. Vanhecke: Geodesic spheres and naturally reductive homogeneous spaces. *Riv. Mat. Univ. Parma (4)* **10** (1984), 123-131.

[TrVa3] F. Tricerri, L. Vanhecke: Geometry of a class of non-symmetric harmonic manifolds. *Differential Geometry and Its Applications*, Proc. Conf. Opava/Czechoslovakia 1992 (Eds. O. Kowalski and D. Krupka), Silesian University, Opava, 1993, 415-426.

[TrVa4] F. Tricerri, L. Vanhecke: The geometry of the Damek-Ricci harmonic spaces. In preparation.

[Van1] L. Vanhecke: Some solved and unsolved problems about harmonic and commutative spaces. *Bull. Soc. Math. Belg. Sér. B* **34** (1982), 1-24.

[Van2] L. Vanhecke: Geometry in normal and tubular neighborhoods. Proc. Workshop on Differential Geometry and Topology Cala Gonone (Sardinia) 1988, *Rend. Sem. Fac. Sci. Univ. Cagliari*, Supplemento al vol. **58** (1988), 73-176.

[VaWi1] L. Vanhecke, T.J. Willmore: Riemann extensions of D'Atri spaces. *Tensor (N.S.)* **38** (1982), 154-158.

[VaWi2] L. Vanhecke, T.J. Willmore: Interaction of tubes and spheres. *Math. Ann.* **263** (1983), 31-42.

[Wan1] Q.M. Wang: Isoparametric hypersurfaces in complex projective spaces. *Differential geometry and differential equations*, Proc. 1980 Beijing Symp. 1982, 1509-1523.

[Wan2] Q.M. Wang: Isoparametric functions on Riemannian manifolds I. *Math. Ann.* **277** (1987), 639-646.

[Wil] T.J. Willmore: *Riemannian geometry.* Oxford Science Publications, Clarendon Press, Oxford, 1993.

[Wol1] T. Wolter: Einstein metrics on solvable groups. *Math. Z.* **206** (1991), 457-471.

[Wol2] T. Wolter: Geometry of homogeneous Hadamard manifolds. *Internat. J. Math. Math. Sci.* **2** (1991), 223-234.

[Zil1] W. Ziller: The Jacobi equation on naturally reductive compact Riemannian homogeneous spaces. *Comment. Math. Helv.* **52** (1977), 573-590.

[Zil2] W. Ziller: Homogeneous Einstein metrics on spheres and projective spaces. *Math. Ann.* **259** (1982), 351-358.

Index

Vol. 1502: C. Simpson, Asymptotic Behavior of Monodromy. V, 139 pages. 1991.

Vol. 1503: S. Shokranian, The Selberg-Arthur Trace Formula (Lectures by J. Arthur). VII, 97 pages. 1991.

Vol. 1504: J. Cheeger, M. Gromov, C. Okonek, P. Pansu, Geometric Topology: Recent Developments. Editors: P. de Bartolomeis, F. Tricerri. VII, 197 pages. 1991.

Vol. 1505: K. Kajitani, T. Nishitani, The Hyperbolic Cauchy Problem. VII, 168 pages. 1991.

Vol. 1506: A. Buium, Differential Algebraic Groups of Finite Dimension. XV, 145 pages. 1992.

Vol. 1507: K. Hulek, T. Peternell, M. Schneider, F.-O. Schreyer (Eds.), Complex Algebraic Varieties. Proceedings, 1990. VII, 179 pages. 1992.

Vol. 1508: M. Vuorinen (Ed.), Quasiconformal Space Mappings. A Collection of Surveys 1960-1990. IX, 148 pages. 1992.

Vol. 1509: J. Aguadé, M. Castellet, F. R. Cohen (Eds.), Algebraic Topology - Homotopy and Group Cohomology. Proceedings, 1990. X, 330 pages. 1992.

Vol. 1510: P. P. Kulish (Ed.), Quantum Groups. Proceedings, 1990. XII, 398 pages. 1992.

Vol. 1511: B. S. Yadav, D. Singh (Eds.), Functional Analysis and Operator Theory. Proceedings, 1990. VIII, 223 pages. 1992.

Vol. 1512: L. M. Adleman, M.-D. A. Huang, Primality Testing and Abelian Varieties Over Finite Fields. VII, 142 pages. 1992.

Vol. 1513: L. S. Block, W. A. Coppel, Dynamics in One Dimension. VIII, 249 pages. 1992.

Vol. 1514: U. Krengel, K. Richter, V. Warstat (Eds.), Ergodic Theory and Related Topics III, Proceedings, 1990. VIII, 236 pages. 1992.

Vol. 1515: E. Ballico, F. Catanese, C. Ciliberto (Eds.), Classification of Irregular Varieties. Proceedings, 1990. VII, 149 pages. 1992.

Vol. 1516: R. A. Lorentz, Multivariate Birkhoff Interpolation. IX, 192 pages. 1992.

Vol. 1517: K. Keimel, W. Roth, Ordered Cones and Approximation. VI, 134 pages. 1992.

Vol. 1518: H. Stichtenoth, M. A. Tsfasman (Eds.), Coding Theory and Algebraic Geometry. Proceedings, 1991. VIII, 223 pages. 1992.

Vol. 1519: M. W. Short, The Primitive Soluble Permutation Groups of Degree less than 256. IX, 145 pages. 1992.

Vol. 1520: Yu. G. Borisovich, Yu. E. Gliklikh (Eds.), Global Analysis – Studies and Applications V. VII, 284 pages. 1992.

Vol. 1521: S. Busenberg, B. Forte, H. K. Kuiken, Mathematical Modelling of Industrial Process. Bari, 1990. Editors: V. Capasso, A. Fasano. VII, 162 pages. 1992.

Vol. 1522: J.-M. Delort, F. B. I. Transformation. VII, 101 pages. 1992.

Vol. 1523: W. Xue, Rings with Morita Duality. X, 168 pages. 1992.

Vol. 1524: M. Coste, L. Mahé, M.-F. Roy (Eds.), Real Algebraic Geometry. Proceedings, 1991. VIII, 418 pages. 1992.

Vol. 1525: C. Casacuberta, M. Castellet (Eds.), Mathematical Research Today and Tomorrow. VII, 112 pages. 1992.

Vol. 1526: J. Azéma, P. A. Meyer, M. Yor (Eds.), Séminaire de Probabilités XXVI. X, 633 pages. 1992.

Vol. 1527: M. I. Freidlin, J.-F. Le Gall, Ecole d'Eté de Probabilités de Saint-Flour XX – 1990. Editor: P. L. Hennequin. VIII, 244 pages. 1992.

Vol. 1528: G. Isac, Complementarity Problems. VI, 297 pages. 1992.

Vol. 1529: J. van Neerven, The Adjoint of a Semigroup of Linear Operators. X, 195 pages. 1992.

Vol. 1530: J. G. Heywood, K. Masuda, R. Rautmann, S. A. Solonnikov (Eds.), The Navier-Stokes Equations II – Theory and Numerical Methods. IX, 322 pages. 1992.

Vol. 1531: M. Stoer, Design of Survivable Networks. IV, 206 pages. 1992.

Vol. 1532: J. F. Colombeau, Multiplication of Distributions. X, 184 pages. 1992.

Vol. 1533: P. Jipsen, H. Rose, Varieties of Lattices. X, 162 pages. 1992.

Vol. 1534: C. Greither, Cyclic Galois Extensions of Commutative Rings. X, 145 pages. 1992.

Vol. 1535: A. B. Evans, Orthomorphism Graphs of Groups. VIII, 114 pages. 1992.

Vol. 1536: M. K. Kwong, A. Zettl, Norm Inequalities for Derivatives and Differences. VII, 150 pages. 1992.

Vol. 1537: P. Fitzpatrick, M. Martelli, J. Mawhin, R. Nussbaum, Topological Methods for Ordinary Differential Equations. Montecatini Terme, 1991. Editors: M. Furi, P. Zecca. VII, 218 pages. 1993.

Vol. 1538: P.-A. Meyer, Quantum Probability for Probabilists. X, 287 pages. 1993.

Vol. 1539: M. Coornaert, A. Papadopoulos, Symbolic Dynamics and Hyperbolic Groups. VIII, 138 pages. 1993.

Vol. 1540: H. Komatsu (Ed.), Functional Analysis and Related Topics, 1991. Proceedings. XXI, 413 pages. 1993.

Vol. 1541: D. A. Dawson, B. Maisonneuve, J. Spencer, Ecole d' Eté de Probabilités de Saint-Flour XXI - 1991. Editor: P. L. Hennequin. VIII, 356 pages. 1993.

Vol. 1542: J.Fröhlich, Th.Kerler, Quantum Groups, Quantum Categories and Quantum Field Theory. VII, 431 pages. 1993.

Vol. 1543: A. L. Dontchev, T. Zolezzi, Well-Posed Optimization Problems. XII, 421 pages. 1993.

Vol. 1544: M.Schürmann, White Noise on Bialgebras. VII, 146 pages. 1993.

Vol. 1545: J. Morgan, K. O'Grady, Differential Topology of Complex Surfaces. VIII, 224 pages. 1993.

Vol. 1546: V. V. Kalashnikov, V. M. Zolotarev (Eds.), Stability Problems for Stochastic Models. Proceedings, 1991. VIII, 229 pages. 1993.

Vol. 1547: P. Harmand, D. Werner, W. Werner, M-ideals in Banach Spaces and Banach Algebras. VIII, 387 pages. 1993.

Vol. 1548: T. Urabe, Dynkin Graphs and Quadrilateral Singularities. VI, 233 pages. 1993.

Vol. 1549: G. Vainikko, Multidimensional Weakly Singular Integral Equations. XI, 159 pages. 1993.

Vol. 1550: A. A. Gonchar, E. B. Saff (Eds.), Methods of Approximation Theory in Complex Analysis and Mathematical Physics IV, 222 pages, 1993.

Vol. 1551: L. Arkeryd, P. L. Lions, P.A. Markowich, S.R. S. Varadhan. Nonequilibrium Problems in Many-Particle Systems. Montecatini, 1992. Editors: C. Cercignani, M. Pulvirenti. VII, 158 pages 1993.

Vol. 1552: J. Hilgert, K.-H. Neeb, Lie Semigroups and their Applications. XII, 315 pages. 1993.

Vol. 1553: J.-L- Colliot-Thélène, J. Kato, P. Vojta. Arithmetic Algebraic Geometry. Trento, 1991. Editor: E. Ballico. VII, 223 pages. 1993.

Vol. 1554: A. K. Lenstra, H. W. Lenstra, Jr. (Eds.), The Development of the Number Field Sieve. VIII, 131 pages. 1993.

Vol. 1555: O. Liess, Conical Refraction and Higher Microlocalization. X, 389 pages. 1993.

Vol. 1556: S. B. Kuksin, Nearly Integrable Infinite-Dimensional Hamiltonian Systems. XXVII, 101 pages. 1993.

Vol. 1557: J. Azéma, P. A. Meyer, M. Yor (Eds.), Séminaire de Probabilités XXVII. VI, 327 pages. 1993.

Vol. 1558: T. J. Bridges, J. E. Furter, Singularity Theory and Equivariant Symplectic Maps. VI, 226 pages. 1993.

Vol. 1559: V. G. Sprindžuk, Classical Diophantine Equations. XII, 228 pages. 1993.

Vol. 1560: T. Bartsch, Topological Methods for Variational Problems with Symmetries. X, 152 pages. 1993.

Vol. 1561: I. S. Molchanov, Limit Theorems for Unions of Random Closed Sets. X, 157 pages. 1993.

Vol. 1562: G. Harder, Eisensteinkohomologie und die Konstruktion gemischter Motive. XX, 184 pages. 1993.

Vol. 1563: E. Fabes, M. Fukushima, L. Gross, C. Kenig, M. Röckner, D. W. Stroock, Dirichlet Forms. Varenna, 1992. Editors: G. Dell'Antonio, U. Mosco. VII, 245 pages. 1993.

Vol. 1564: J. Jorgenson, S. Lang, Basic Analysis of Regularized Series and Products. IX, 122 pages. 1993.

Vol. 1565: L. Boutet de Monvel, C. De Concini, C. Procesi, P. Schapira, M. Vergne. D-modules, Representation Theory, and Quantum Groups. Venezia, 1992. Editors: G. Zampieri, A. D'Agnolo. VII, 217 pages. 1993.

Vol. 1566: B. Edixhoven, J.-H. Evertse (Eds.), Diophantine Approximation and Abelian Varieties. XIII, 127 pages. 1993.

Vol. 1567: R. L. Dobrushin, S. Kusuoka, Statistical Mechanics and Fractals. VII, 98 pages. 1993.

Vol. 1568: F. Weisz, Martingale Hardy Spaces and their Application in Fourier Analysis. VIII, 217 pages. 1994.

Vol. 1569: V. Totik, Weighted Approximation with Varying Weight. VI, 117 pages. 1994.

Vol. 1570: R. deLaubenfels, Existence Families, Functional Calculi and Evolution Equations. XV, 234 pages. 1994.

Vol. 1571: S. Yu. Pilyugin, The Space of Dynamical Systems with the C^0-Topology. X, 188 pages. 1994.

Vol. 1572: L. Göttsche, Hilbert Schemes of Zero-Dimensional Subschemes of Smooth Varieties. IX, 196 pages. 1994.

Vol. 1573: V. P. Havin, N. K. Nikolski (Eds.), Linear and Complex Analysis – Problem Book 3 – Part I. XXII, 489 pages. 1994.

Vol. 1574: V. P. Havin, N. K. Nikolski (Eds.), Linear and Complex Analysis – Problem Book 3 – Part II. XXII, 507 pages. 1994.

Vol. 1575: M. Mitrea, Clifford Wavelets, Singular Integrals, and Hardy Spaces. XI, 116 pages. 1994.

Vol. 1576: K. Kitahara, Spaces of Approximating Functions with Haar-Like Conditions. X, 110 pages. 1994.

Vol. 1577: N. Obata, White Noise Calculus and Fock Space. X, 183 pages. 1994.

Vol. 1578: J. Bernstein, V. Lunts, Equivariant Sheaves and Functors. V, 139 pages. 1994.

Vol. 1579: N. Kazamaki, Continuous Exponential Martingales and *BMO*. VII, 91 pages. 1994.

Vol. 1580: M. Milman, Extrapolation and Optimal Decompositions with Applications to Analysis. XI, 161 pages. 1994.

Vol. 1581: D. Bakry, R. D. Gill, S. A. Molchanov, Lectures on Probability Theory. Editor: P. Bernard. VIII, 420 pages. 1994.

Vol. 1582: W. Balser, From Divergent Power Series to Analytic Functions. X, 108 pages. 1994.

Vol. 1583: J. Azéma, P. A. Meyer, M. Yor (Eds.), Séminaire de Probabilités XXVIII. VI, 334 pages. 1994.

Vol. 1584: M. Brokate, N. Kenmochi, I. Müller, J. F. Rodriguez, C. Verdi, Phase Transitions and Hysteresis. Montecatini Terme, 1993. Editor: A. Visintin. VII. 291 pages. 1994.

Vol. 1585: G. Frey (Ed.), On Artin's Conjecture for Odd 2-dimensional Representations. VIII, 148 pages. 1994.

Vol. 1586: R. Nillsen, Difference Spaces and Invariant Linear Forms. XII, 186 pages. 1994.

Vol. 1587: N. Xi, Representations of Affine Hecke Algebras. VIII, 137 pages. 1994.

Vol. 1588: C. Scheiderer, Real and Étale Cohomology. XXIV, 273 pages. 1994.

Vol. 1589: J. Bellissard, M. Degli Esposti, G. Forni, S. Graffi, S. Isola, J. N. Mather, Transition to Chaos in Classical and Quantum Mechanics. Montecatini, 1991. Editor: S. Graffi. VII, 192 pages. 1994.

Vol. 1590: P. M. Soardi, Potential Theory on Infinite Networks. VIII, 187 pages. 1994.

Vol. 1591: M. Abate, G. Patrizio, Finsler Metrics – A Global Approach. IX, 180 pages. 1994.

Vol. 1592: K. W. Breitung, Asymptotic Approximations for Probability Integrals. IX, 146 pages. 1994.

Vol. 1593: J. Jorgenson & S. Lang, D. Goldfeld, Explicit Formulas for Regularized Products and Series. VIII, 154 pages. 1994.

Vol. 1594: M. Green, J. Murre, C. Voisin, Algebraic Cycles and Hodge Theory. Torino, 1993. Editors: A. Albano, F. Bardelli. VII, 275 pages. 1994.

Vol. 1595: R.D.M. Accola, Topics in the Theory of Riemann Surfaces. IX, 105 pages. 1994.

Vol. 1596: L. Heindorf, L. B. Shapiro, Nearly Projective Boolean Algebras. X, 202 pages. 1994.

Vol. 1597: B. Herzog, Kodaira-Spencer Maps in Local Algebra. XVII, 176 pages. 1994.

Vol. 1598: J. Berndt, F. Tricerri, L. Vanhecke, Generalized Heisenberg Groups and Damek-Ricci Harmonic Spaces. VIII, 125 pages. 1995.